Diverse Topics in
Science and Technology

Diverse Topics in Science and Technology

Benu Chatterjee

authorHOUSE®

AuthorHouse™ UK Ltd.
1663 Liberty Drive
Bloomington, IN 47403 USA
www.authorhouse.co.uk
Phone: 0800.197.4150

Published by AuthorHouse 09/26/2013

ISBN: 978-1-4918-7755-5 (sc)
ISBN: 978-1-4918-7615-2 (hc)
ISBN: 978-1-4918-7756-2 (e)

List of Articles

Part 1: Science and Nature

Part 2: Materials and Engineering

"To my wife Eileen for her incredible patience with me".

Preface

The book provides comprehensive reviews of several aspects of science, engineering and technology. The basic concepts of the three faculties are intricately linked. Science starts with a physical system and seeks to develop a scientific theory. Engineering is the application of science for human benefits, while technology utilizes scientific and engineering knowledge to achieve a practical result. The initial concept of writing the book was inspired by television documentaries, books, magazines etc. supplemented by the spirit of free thinking.

In view of multi facets of interest in the life of a scientific community, it is difficult to choose specific topics. Nevertheless, subjects are selected over a range of ideas reflecting author's curiosities in science, engineering and technology. The book is written in an endeavour to emphasize the importance of understanding the mechanics of how science and engineering, and also scientists and engineers function in our increasingly technological world since the early days of ancient humans.

Eighteen comprehensive reviews are presented in two parts. The first, covering chapters 1 to 13, is based on science and nature. Articles include fascinating story of super active Earth's core on which we are standing, environmental issues, origin of different mineral colours, and science of both common experiences and natural phenomena.

Five articles are presented on materials and technology in the second part which begins with a review of the basic strategy of metaphorically speaking green materials. This is important in view of the growing awareness of issues on environmental protection of everyday life in the field of materials science and engineering. The next article is on the incredible metallurgical expertise of ancient humans. Their rudimentary techniques were certainly good enough to get the job done, and surprisingly form the basis of some of our modern metal processing methods. Subsequent articles review the development of materials and engineering technology from ancient to medieval to modern time with an eventual outlook for the future.

Articles are written in simple terms to stimulate, inspire the grey cells of the readers who are enthusiastic about science and technology. Some of the articles are based on the basic knowledge of physics and mathematics that would provide a comprehensive tour for the physical scientists in the realm of several common everyday experiences and natural events.

Information on further details including sources/further reading of all articles can be obtained from "benu-chatterjee.suite.101.com" and "science-to-go.weebly.com".

PART - 1

Science and Nature

Super Active Earth's Core beneath Our Feet

It is incredible to think that although we walk on nice cool ground of Earth's crust, there is a geological concert going on underneath our feet perpetually orchestrated by massive forces and immense transfer of heat where molten metals flow like water. Following a brief account of Earth's interior, present aim is mainly to discuss some of the interesting enigmas of Earth's core.

Study of Earth's Interior

Human's eternal urge to explore the unknowns such as landing on the Moon or sending satellites to various solar planets, continues with our curiosity about Earth's interior.

Scientific understanding of Earth's inside as layered spherical shells, started from an early part of the 20th century. Seismology provides the best resolution technology of all geophysical probes to map out Earth's structure and composition. Seismic waves go through Earth's interior when an earthquake occurs. For example, Tohoku earthquake in Japan in 2011 transmitted seismic waves through Earth's centre to Chile located on the other side of the globe.

Other methods of investigating Earth's interior include studies of gravitational and magnetic fields, and surface topography of rocks from Earth's interior that reached the surface via volcanic eruption.

1). Seismology and Earth's interior

Transmission of seismic waves occurs via reflection and refraction (bending) of the waves at discontinuities and gradients they encounter. Changes in the material properties of Earth's interior such as composition, mineral phase and packing structure as well as the temperature and pressure directly affect the nominal velocity (~10 km per second) of seismic wave as it passes through Earth's centre.

2). Depth of Earth's interior layers

Based on changes in the velocity of seismic waves, Earth's layered interior is estimated to consist of an outer solid crust of silicate directly beneath our feet, followed by a highly viscous mantle, a less viscous liquid outer core and a solid inner core. The characteristics of different spherical layers are as follows in increasing order of distance from the surface:

crust (30 km) of solid outer shell on which we live at pressure ~1 atmosphere or 0.0001 GPa and normal air temperature

mantle (upper: 720 km and lower: 2,170 km) made up of semi-molten rock, called magma at ~24 GPa and ~ 1,600 C

outer core (2,260 km) of molten iron and iron-(~5%)nickel alloy at 136 GPa and 3,700 C, and

inner core (1,220 km) of solid mass of iron and nickel at 364 GPa and 5,500 C

Study of Earth's core

Earth was formed around five billion years ago via massive conglomeration and high-velocity bombardment of meteorites and comets. An immense amount of heat was generated. Heavier molten materials like iron and iron-nickel alloy from the meteorites sank into Earth's core, while lighter silicates, other oxygen compounds and water from comets rose near the surface.

Extreme heat and pressure have made Earth's core as one of the most inhospitable places in the entire solar system. The inner solid core is smaller than Moon, while Mars could snugly fit inside the outer core.

1). Inner core

A crushing pressure of ~3.64 million atmospheres (364 GPa) compared to 0.0001 GPa on Earth's surface is exerted on the inner core by gravity from above. As a result, the inner core, mostly iron, remains solid despite being as hot (> 5000 C) as Sun's surface.

Seismologist proved the inner core to be solid by trying to detect transverse shear seismic S-wave which can only penetrate through a solid. However, S-wave is very small to detect. One would, therefore, need a huge amount of data to sift through. Dr. Arwen Deuss of Cambridge University just did that by collecting seismic data from around the globe (47 seismic stations), and eventually observed the tiny S-wave travelling through the inner core, proving it to be solid.

2). Outer core

Unlike the inner core, the outer core is not under enough high pressure to be solid. It consists of molten metals of iron and nickel whose dynamic activity is similar to that happening in the mixing of upper and lower mantle. The turbulence caused by a convection process in the liquid metals, as discussed later on, is believed to create Earth's magnetic field. The average strength of the magnetic field is highest (25 gauss) inside the outer core, which gradually decreases with distance from the core, and eventually end up ~ 50 times weaker at the surface of Earth.

Enigmas of Earth's core

Unexplained crystallography at or near Earth's core

1). Mineral structure of D" layer

The nature of mineral at the boundary between lower mantle and outer core, the so-called D" layer of ~ 300 km thickness at a depth of 2890 km where pressure and temperature are around 125 GPa and 2,200 C respectively, is unclear. The D" layer is expected to be a magnesium silicate mineral with perovskite crystallographic structure. However, seismic data revealed anomalous structure with relative abundance of magnesium silicates.

2). Structure of iron in the inner core

Compressed seismic P-waves which can go through both solid and liquid, are known to travel faster through the inner core by about 3% or ~5 seconds in the direction of Earth's polar axis (North to South) compared to that along the equatorial plane (East to West). This phenomenon is termed as seismic anisotropy. The anomaly in seismic speeds can be resolved if the inner core, mostly of iron, can be proved to have a texture with the "fast axis" of the crystals mostly oriented in the North-South direction.

Earth's mysterious magnetic phenomena

1). Importance of Earth's magnetic field for our survival

Scientists are fascinated by the fact that Earth's outer core is responsible for Earth's magnetic i.e. geomagnetic field which is vital to the evolution of life on Earth. The magnetic bubble surrounding the Earth called magnetosphere contains the necessary magnetic field to deflect away harmful plasma storms and deadly charged particles from solar radiation. This action protects us from being fired out of existence. Gradual perish of atmosphere on Mars, for example, is due to loss of magnetic field which has probably led the Red Planet to become a dead world.

Our insufficient knowledge on the creation of planetary magnetism has left us pondering about the unanswered questions on the mechanics of origin and survival for billions of years of Earth's magnetic field. It is even more puzzling that the smallest planet Mercury has a magnetic field, while both Mars and Venus have none.

2). Criteria of geomagnetic field source

Earth's magnetic i.e. geomagnetic field must be continuously fed with energy for its survival over billions of years. Life on Earth would, otherwise, decay and disappear by interacting with the solar radiation. Mechanisms are sought to explain perpetual generation of geomagnetic field, and an energy source to power it all the time.

3). Possible mechanisms of geomagnetism

The traditional belief that a giant magnet lying between the north and south poles inside the Earth creates geomagnetism is a myth. This is because the temperature inside the Earth is far above the Curie point of 770 C for iron, making it impossible for the core to remain magnetized.

A most plausible alternative mechanism similar to a dynamo operation is briefly discussed next whereby mechanical energy is converted into electrical energy.

4). Geodynamo theory of Earth's magnetism (geomagnetism)

Most scientists agree that Earth's magnetic field arises from the turbulence of liquid metal of mostly iron in the outer core. This movement originates from convection of heat left over from the birth of the Earth.

Molten iron conducts electricity which, as expected, is surrounded by magnetic fields. Geomagnetism is created by the circulating electric currents in Earth's molten outer core via self-sustaining geodynamo which generates magnetic fields assisted by a "seed" field. The prerequisites for geodynamo can be summarised as

a) fluid with good electrical conductivity, namely molten iron or iron-nickel alloy.
b) high temperature to keep the metal in liquid state,
c) enough energy difference in terms of temperature and pressure to provide convection movement of the molten metal with sufficient speed and appropriate flow pattern assisted by Coriolis Force from Earth's rotation, and finally
d) presence of "seed" magnetic field such as provided by the Sun to start the process.

All these conditions are met in the outer core where the temperature is ~4000 C. The turbulence by convection of electrically conductive liquid iron generates magnetic field. Also, electric current via magnetic induction process is generated from Sun's "seed" magnetic field. This newly created electric current in turn produces a magnetic field that interacts with the fluid motion to create a secondary magnetic field. Together, the two fields are stronger than the original. Thus as long as there is sufficient liquid turbulence in the outer core, cycle continues establishing a self-sustaining loop of geodynamo, and hence perpetual generation of Earth's magnetic field with one pole up in Canada and another down in Antarctica.

5). Magnetic flip/reversal

Scientists believe that over the past 400 years, there has been a steady decline in Earth's magnetic field, probably due to changes in the churning motion of the liquid outer core. Since the first measurement reported some 180 years ago, magnetic field strength has waned 10-15%. This would lead to shrinkage of magnetosphere and hence less shielding from harmful solar radiation. There are already little bit of the radiation leaking through near the poles that causes the aurora.

There had been reports of delicate sensors in Hubble space telescope, satellites and computer in space shuttles malfunctioning when passing over the Atlantic. It soon became clear that this type of incidents dubbed as "South Atlantic Anomaly" (SAA) was tightly clustered around the centre of South America and South Atlantic. The region of anomaly is growing all the time, and it is speculated that in just over 200 years, the region could cover the entire Southern Hemisphere.

SAA is suspected to be much more than just an inconvenience to satellite operators. It may be the first indication of a profound change in Earth's magnetic field. On mapping out Earth's magnetic field, scientists have discovered that the magnetic field under SAA patches has actually flipped with magnetic north pointing south and vice versa. If the patches continue to deepen and spread, it is just a hunch that Earth's entire magnetic field could reach a tipping point and flip/ reverse.

Possible journey to the centre of Earth

Fascinated by the mysteries surrounding Earth's interior, scientists for centuries have been dreaming of reaching Earth's centre some 6400 km (~ 4000 miles) beneath our feet. The deepest depth so far managed by drilling through Earth's crust is 12 km at Kola Superdeep Borehole in Russia, which is merely 0.2% of the distance to the centre of the Earth, and only 40% of the thickness of Earth's crust.

Temperature at the bottom of the Borehole was already ~ 180C which would extend to about 4000 C along with the pressure reaching very high on further descending down. Drilling through these extreme hostile conditions, much worse than on the Sun, is simply not possible, unlike Jules Verne's imaginary story.

However, this set back did not deter scientists to consider alternative routes. Simulation programmes are developed to investigate the conditions of Earth's core in the laboratory.

Laboratory simulation of the extreme environment of Earth's core

1). Resolution of D" layer structure

The research team of Prof. Kei Hirose of Tokyo Inst. of Technology succeeded the seemingly impossible task of reaching the extreme conditions (i.e. 125 GPa and over 2200C) of the D" layer using a laser-heated diamond-anvil cell. A sample (25µm thick) of magnesium silicate was sandwiched between the flattened tips of two opposing diamond anvils. The crystal structure was studied in-situ by Synchrotron x-ray diffraction method.

An unknown structure was observed with ~ 1% higher density than the expected perovskite structure. Dubbed as "postperovskite" mineral, this new material can explain the puzzle about the type of mineral in D" layer. Postperovskite has crystal structure similar to mica. Its electrical conductivity is about four orders of magnitude higher than the corresponding perovskite form.

2). Creation of inner core conditions

Flushed with success, Hirose continued to push harder to attain conditions of the inner core. He succeeded in pressing an iron sample to as high as 377 GPa at over 5400 C. Results showed that at these conditions similar to inner core, iron has a hexagonal closed-packed (HCP) structure with the atoms bonded at high density.

Speculation is that these HCP iron crystals in the core are huge up to 10 km in height. They are preferentially aligned like a forest with the c-axis parallel to Earth's rotational axis. This unidirectional alignment can explain the observed seismic anisotropy of the inner core mentioned earlier.

Laboratory study of Earth's magnetic field

1). Possibility of creating geodynamo in the lab

Scientists try to develop a system mimicking Earth's dynamo. But the problem lies with the difficulty of modelling down the immense size of Earth's core, and also the heat or the speed at which it spins. For example, a 100-metre sphere filled with liquid metal would have to rotate insanely fast at more than 100,000 revolutions per minute (or 100,000 miles per hour at its equator) to achieve true dynamic similarity with Earth's core. An estimated 100 kilowatts of thermal power would be needed to drive the kind of turbulence expected in Earth's core. A lesser rate of revolution would need more thermal power to drive the necessary degree of convection. Undaunted by the reality of the whole thing being beyond normal comprehension, scientists are still trying to create geodynamo in the laboratory.

It is universally agreed that since iron melts at high temperature (1538 C), an alternative metal such as sodium with melting point at 97.7 C would be a better candidate to manage in the lab as the working fluid. Liquid sodium, like iron, strongly interacts with magnetic field.

For container designs, the best suite of sodium experiments uses spherical shapes to mimic the round shape of Earth's geodynamo. Prof. Lathrop and his team at the University of Maryland, have thus designed two stainless steel spheres nestled one within the other. The inner sphere of 1 metre across represents Earth's solid inner core, while a 3-metre tall outer sphere represents outer core.

The space between the two is filled with 12 tons of liquid sodium mimicking the liquid iron in Earth's outer core. It is hoped that spinning independently at 4 and 12 revolutions per second for the outer and inner spheres respectively, will create its own dynamo to provide a self-sustaining magnetic field with Earth's natural magnetism as a "seed field" to kick start the process. The world awaits the results with great anticipation once the machine is powered in near future.

New alternative design using superheated state of matter known as plasma instead of liquid sodium is being developed at Wisconsin University, Madison. This would upgrade the study of geodynamo by several notches.

2). Thoughts on magnetic reversals

Earth's magnetic field has never reversed while people are around to record it. Studies of rock samples definitely point out magnetic reversal occurring many hundreds of times over the past billion years. The last time reversal took place was 780,000 years ago when Homo erectus was still learning how to make stone tools.

It appears that every 100,000 to million years, the north-south orientation of the magnetosphere reverses, and is often preceded by an overall weakening of magnetic field. Based on present detection of SAA along with the report of a steady decline in field strength, one can speculate that we are heading for a magnetic reversal.

Such magnetic flip, even if imminent, is not something that would occur overnight. Magnetic reversal can take hundreds or thousands of years to mature from start to finish. During this period one can only speculate that the magnetic field would be pretty confused with perhaps magnetic poles wandering to the equator taking with them the spectacular Northern Lights.

Final comments

After dreaming of reaching the centre of the Earth, scientists are now uncovering a bizarre and alien world via simulation programmes. Unlike anything else we experience on Earth's surface, the extreme conditions in the interior are unique in the solar system. Inside the Earth, the outer core of the size of Mars has perpetual violent storms in a sea of molten iron, while the solid inner core of the size of the Moon consists of giant forest of iron crystals. A unique magnetosphere created by the outer core, protects us from burning out by solar radiation. With all this, one can envisage Earth's interior as a planet buried within a planet of Earth we know.

Basics of Aerosols:
Sources, Sizes and Impacts

The general features of hazardous aerosols defined as ubiquitous small particles of solid or liquid suspended in the Earth's atmosphere are reviewed.

The word "aerosols" may prompt people to think about the high-pressure spray cans containing a liquefied gas which propels products like deodorants, air fresheners and cleaning sprays through a valve. However, in the present context the object has been to review the basics of airborne tiny particles of aerosols in terms of their origins, sizes and impact on humans.

Air can be contaminated with particles ranging from dust, pollen, soot, smoke, to liquid droplets. On taking a deep breath, even if the air looks clear, one is likely to inhale aerosols or particulate matters (PM) in the form of tens of millions of small specs (sub-micrometer to several micrometers, µm) of airborne solid particles and liquid droplets. These tiny particles can have a major impact on health and the climate.

Natural Sources of Aerosols

Bulk (80—90% by mass) of the aerosols is emitted directly (primary type) as particles into the atmosphere by natural processes, such as wind lifting of dust particles, sea spray, volcanoes, combustion during biomass burning etc.

Sea salt and dust are the two most abundant aerosols and produce particles larger than from other sources. Wind-driven spray from ocean waves throws sea salt aloft, and sandstorms whip small pieces of mineral dust from deserts into the atmosphere. Volcanoes eject huge amounts of ash into the air, as well as sulphur dioxide and other gases yielding sulphates as solid particles. Forest fires result in partially burned organic carbon compounds. Also, certain plants generate gases that on being airborne react with other substances to produce aerosols, such as "smoke" in the Great Smoky Mountains of the United States.

Man-made (Anthropogenic) Sources of Aerosols

Anthropogenic activity such as burning fossil fuels in vehicles, power plants, combustion of coal and incinerators produces 10-20 % of global aerosols indirectly (secondary type) via atmospheric chemical conversion of gases such as sulphur dioxide, nitrogen oxides and ammonia to particles of sulphates, nitrates and ammonium compounds. Industrial processes like smelting are prolific producers of sulphates, nitrates, black carbon (soot) and other particles.

Man-made sources, though less abundant than natural sources, dominate the air downwind of urban and industrial areas. Alteration of land surface by deforestation, overgrazing and drought would enhance rate of entry of dust aerosols into the atmosphere.

A look at the location of aerosol plumes would enable one to discriminate sources. Natural aerosols, like salt particles from sea spray, are typically widespread over large areas, and unlike man-made sources, they are not particularly concentrated downwind of urban areas.

Also, particles produced by human activity are generally much smaller than those of natural origins.

Background to Sizes of Aerosol Particles

The aerodynamic properties of aerosol particles determine how they are transported in air and how they can be removed from it. Since particles can have irregular shapes, their aerodynamic behaviour is expressed in terms of the diameter of an idealised sphere. This is known as aerodynamic diameter or particle size. These sizes range from a few nanometres, nm (less than the width of the smallest viruses) to several tens of μm (about the diameter of a human hair).

Categorization of Aerosol Particle Sizes

Toxicologists categorize particles as fractions, such as thoracic ≤ (equal or smaller than) 10 μm, fine ≤2.5 μm and ultrafine ≤0.1 μm fractions. Regulatory agencies as well as meteorologists typically call them PM and express as total mass of aerosols per unit volume. For example, PM-10 refers to the mass of aerosol particles with a diameter of 10 μm or less.

Effectiveness

Particles with diameter ≤ 0.05 μm disappear quickly by agglomerating with other particles. However, particles between 0.1 μm and 2.5 μm persist for a long period of time—up to 2 weeks—and pose the biggest threat to human health because they are small enough to be inhaled deep into the lungs.

Particle pollution can also occur for the range 2.5 μm-10 μm. In contrast, particles >10 μm in diameter are too heavy to stay suspended and consequently, they settle to the ground by gravity in matter of hours.

Global Distribution of Aerosol Particles

A global map on average distribution of fine aerosol particles indicates highest concentrations of PM-2.5 over highly industrialized areas in eastern Asia. The high concentrations over northern Africa and the Middle East are likely to be due to the fine dust from the deserts.

Based on data representing typical model distribution of aerosol concentration, urban environments are characterized by higher concentrations of aerosol particles (as much as 100,000 particles per cm3) than in maritime atmosphere (~100 particles per cm3). Particle concentration in rural areas is lower compared to urban aerosols. The maximum concentrations in both urban and rural atmosphere are reached at PM-0.01, while it is at PM-0.1 in maritime climate.

Health Impact and Effects of Aerosols

There have been a large number of deaths and other health problems associated with particulate pollution. The main determinant in affecting the respiratory tract when inhaled is the particle size. It is generally considered that PM-10 or even PM-2.5 fraction can penetrate the deepest part of the lungs.

The smaller particles (e.g., less than 100 nm) can pass through cell membranes and may even be more damaging to the cardiovascular system. Larger particles are, however, generally filtered via the nose and throat. Size aside, further complexity can arise from the shape. For example, it's well-known that the feathery-shape of asbestos can be dire once lodged in the lungs.

Aerosols' Impact on Visibility

Degradation of visibility originates largely from anthropogenic aerosols. It occurs due to the light extinction property of aerosols, which is a collective process of light being scattered (changing the direction of photons' propagation) and absorbed (removing photon from the beam resulting in heating) leading to the obliteration of sunlight. The complex refractive index of aerosol particles affect their light extinction property and thus, the visibility of an area. The level of visibility depends on the size, concentration in the path of the light, composition and water uptake of aerosol particles. The visibility is lowest during the dry months when aerosol dusts are heavy; conditions improve with increased rainfall.

The Environmental Impact of Aerosols

The scattering and absorption of solar and terrestrial radiation by aerosols would affect Earth's climate and the planet's radiation budget.

In an indirect way, aerosols tend to modify cloud properties by acting as cloud condensation nuclei. The formation of cloud droplets is thereby promoted, resulting in cloud formation at lower super-saturation points than if there were no aerosols.

An oxymoron effect is exhibited by aerosols in terms of warming/cooling of the planet. Aerosols can cool the Earth's surface by reflecting sunlight to space and by forcing changes in cloud microphysics that would consequently increase cloud reflection of sunlight. On the other hand, black carbon (soot, from incomplete combustion) absorbs light and it can warm the atmosphere and greenhouse gases like carbon dioxide. It is estimated that the cooling effect cancels about 10% of the global warming effect.

Ozone depletion in stratosphere, and pollution in troposphere are other climate areas of concern. Man-made chemicals, such as the CFC (chlofluorocarbon) family, drift upwards to the stratosphere and they're broken down by strong sunlight. This results in the release of chlorine, which is capable of depleting the natural ozone layer which comprises 90% of Earth's ozone. This "hole" in the ozone layer would allow more of the sun's UV radiation to reach Earth, causing skin cancer and reducing plant and crop growth. Natural causes aside, troposphere air pollution can arise from anthropogenic secondary aerosols and volatile organic compounds.

Impact on Acid Deposition

Acid deposition, commonly known as acid rain, adversely affects forestry, vegetation and fishing industry. The most important acid-forming species are sulphur dioxide and nitrogen oxides which are converted to sulphuric and nitric acids respectively during atmospheric transport. Ammonia gas in the atmosphere can neutralise the acids, but in soil, it's changed into nitric acid by microorganisms. The ultimate result is a devastating impact on vegetation.

Motivated by climate changes and adverse health effects, research on aerosols has been intensified over the past few decades for a better understanding of the mechanisms and factors controlling their chemistry. Examination of global satellite images, along with global scale models, can provide the scientists with the best tools to determine and analyze the effects of aerosols on climate and weather patterns around the world.

Characterization and Origins of Diverse Mineral Colours

The beautiful colours of minerals have been valued in all societies. A comprehensive review of the colour phenomena in minerals is presented.

Examples of colourful minerals

Colour is the most eye-catching feature of minerals. The recognition of colours in minerals goes back to pre-historic ages when charcoal and iron oxides (red hematite mineral) were used to colour cave paintings which still retain their original brightness. The present article is a comprehensive review of the various features of the diverse mineral colours, namely mechanics, characterization, complexities and origins of colours.

Mechanics of Creating Mineral Colours

The visible white light forms a part of the electromagnetic spectrum we can see. It is made up of waves with wavelengths of 350 to 750 nanometres that correspond to the seven colours of the rainbow with violet and red having the shortest and longest wavelengths respectively. Also, wavelength (λ) and frequency (γ) of light wave are related to energy (E) (photon: a basic unit of electromagnetic radiation) as $E = hc/\lambda = h\gamma$ where h is Planck's constant and c is the speed of light. When a solid sample absorbs light, what we see is the sum of the remaining reflected colours that strike our eyes. If all the light energies are reflected back, the sample will appear colourless or white. However, it will appear black if all the energies are absorbed, leaving none of the visible spectrum reaching our eyes.

Minerals develop colours by interacting with white light when only part of the visible spectrum is absorbed leaving the rest to emit. When this happens, the energies absorbed are removed from the emitted light which will no longer be white, but have complementary colour associated with the emitted wavelengths. The observed colour of a mineral thus corresponds to those wave lengths of light that are least absorbed.

Characterization of Mineral Colours—Idiochromatic, Allochromatic and Pseudochromatic

Certain elements termed as chromophores, are pigmenting agents which provide colours to minerals when they are present as a part of the crystal lattice, or as impurity. Chief chromophores are the first row of transition metals in the Periodic Table with atomic numbers 22 to 29, namely titanium, vanadium, chromium, manganese, iron, cobalt, nickel and copper. Mineral colours are classified into three main groups, namely idiochromatic. allochromatic and pseudochromatic.

Idiochramatic ("self coloured") minerals exhibit their own inherent permanent colours that are derived directly from the presence of one or more chromophores as a constituent of the crystal lattice. The property of chromophores determines which wavelengths of light to absorb and which ones to emit. Examples include Iron-based red cinnabar, copper-based blue azurite and green malachite minerals.

Allochromatic ("other coloured") minerals, originally colourless, develop colours owing to the presence of trace amounts (parts per million) of chromophores as impurity in the host crystal lattice. The colourless quartz mineral for example changes to purple colour (amethyst mineral) when iron is present as impurity in the lattice. Similarly, colourless corundum changes to blue sapphire or red ruby by the presence of trace amounts of either iron and titanium, or chromium respectively.

Pseudochromatic ("false coloured") minerals display colour only because of the tricks played by light. These minerals contain layers that create colours by various features of physical optics discussed later on. The colour may vary, but is often a unique property of the mineral. For example, opal, moonstone and labradorite all reflect light in a characteristic way, but the colours are not true to the types of minerals.

Complexities of Mineral Colour and Theories

In mineralogy, colour is one the primary diagnostic features to identify minerals. However, mineral colour is changeable, and unpredictable, and cannot be used for identification. For example, different impurities can cause fluorite mineral to come out in various colours. Also, there are minerals of similar colours. Furthermore, some minerals such as topaz and beryl undergo changes from dull to deeper colour on heating as practised in the gemstone industry. These complex features of mineral colours make it less reliable to identify a mineral from its colour. It is thus important to understand what causes colours in minerals.

In order to show colour, a mineral has to somehow disturb the balance of the light energies with preferential absorption or emission of certain wavelengths/energies of light. The theories of the origins of colours in minerals such as idiochromatic, allochromatic and pseudochromatic which cover most of the familiar minerals, are simple. Delving deeper, Kurt Nassau (The Physics and Chemistry of Colors: The Fifteen Causes of Color, 1983) has separated causes of colour into fifteen mechanisms based on several groups, such as crystal field effects, physical optics, band gap theory and molecular orbital mechanism. However, in order to keep the present article short and simple, only the crystal field theory and physical optics are discussed. Any discussion of either the band gap theory or molecular orbital formalism is excluded—both of which nonetheless relate to less familiar materials such as organic materials, semiconductors and metal conductors.

Crystal Field Theory—Transition Metal Ions and

Colour Centres

Crystal Field Theory is based on two mechanisms involving transition metal ions and colour centres:

Transition metal ions (simple cases) is based on electron interactions, and the crystal field effects can explain both idiochromatic and allochromatic colours. The innermost electrons of chromophore transition metals are too tightly bound to the nucleus to be excited by light energy (photon) because electrons, after all, generally like to move about in pairs. However, valence electrons in these metal ions are unpaired which along with inner 3-d orbitals partially filled with electrons, are prone to excitation to higher energy orbitals by photons. This phenomenon would satisfy the need for pairing up of electrons, and thus becomes responsible for colours in minerals. Excitatation is achieved by absorbing photon from visible incident light. The absorbed energies are subtracted from that of the incident light, resulting in the observed colour.

Transition metal ions (complex cases) lead to characteristics found in, for example, corundum and beryl. Both corundum (aluminium oxide) and beryl (beryllium aluminosilicate) minerals are colourless because all electrons are paired in the crystal lattice allowing no absorption of light. However, addition of same chromium ion ($Cr 3+$) as chromophore impurity to either mineral, changes them to become colourful of different shades, namely red (ruby) and green (emerald) respectively. This phenomenon indicates complexity of crystal field interactions. It is believed that the geometry of the atoms in beryl is different from corundum, allowing the host molecules to interact with chromium more weakly. This results in reduced energy differences between the energy levels which eventually results in developing complementary green colour rather than ruby red in corundum.

The colour centres (F-centres) are defects created by removing atoms usually by radiation from the crystal structure. The resulting "hole" may be filled by an electron from a neighbouring atom. This will leave behind an unpaired electron which, as discussed before, is prone to excitation by photons with ultimate creation of colours. The most familiar examples are amethyst and smoky quartz minerals.

Physical Optics Effects—Scattering, Diffraction, Interference and Dispersion

Colouration of pseudochromatic minerals is best explained by physical optics processes where, unlike crystal field theory, electrons are not involved. Light passing through a mineral interacts with the crystal structure, and colours are produced due to various phenomena such as scattering (which explains blue sky), diffraction (peacock feather), interference (oil slicks) and dispersion (rainbow spectrum through prism) of incident white light.

Scattering occurs due to reflection of light off sub microscopic (the finer the better) particles of solid or liquid. It would cause colours to appear because blue light with shorter wavelength is scattered more than red. As particles become larger, they scatter other colours which join with the blue until the ultimate white colour is reached. Examples include star sapphire, bluish moonstone, and white to bluish colour of some opal. Regularly spaced layers (for diffraction) or thin layers (for interference) of minerals with different indices of refraction will produce colours if the separation of the layers is of the same order of magnitude as the wavelength of light. Opal and feldspars produce flashes of colour via diffraction. The play of colour in labradorite and the tarnish layers on chalcopyrite are good examples of interference colouring. Dispersion is related to mineral's index of refraction. Lights of different wavelengths bend differently when passing obliquely through a mineral, causing light to spread out into colours of the visible spectrum. This is best seen in faceted gemstones such as diamond, pure rutile and zircon.

A remarkable scientific progress has been made in the past two decades in understanding minerals' colouring phenomena. Based on our current knowledge of the optical properties of solids, we nowadays make use, for example, of rubies and sapphires in high power solid-state lasers.

Science of Common Experiences

Girl on a Swing

Based on basic physics and mathematics, scientific explanations are presented for some everyday experiences and natural phenomena.

Background

In the past, some of our ordinary experiences had seldom been accompanied with any explanation of why the facts have arisen. For example, when the use of wheel was discovered, nothing was known of frictional forces, nor any reasons why goods loaded on wheeled vehicles are easier to move than dragging along the ground

Also, sometimes "common sense" was used in an effort to explain facts without testing relevance of the truth. For example, the medicinal properties of herbs like the foxglove as cardiac stimulant was explained in terms of similarity of shape of the flower to human heart.

The desire for systematic and controllable explanations of factual evidence in modern time has generated the knowledge of science. The distinctive goal of science involving physics and mathematics has been to organise our knowledge based on explanatory principles. History shows that physics and mathematics are inextricably linked. In physics, physical environment is studied to explain how things work, while mathematics tends to quantify observations and make predictions.

In a series of articles, five interesting everyday experiences and five natural phenomena are chosen that are explained on the basis of basic science of physics and maths.

Girl on a Swing

Physics

Swings and the flying trapeze are examples of one of the classic topics in physics namely, pendulum. Swings provide a feeling of flying in a controlled manner.

Resonance in swinging

A girl on a swing pushed by her father with constant periodic rather than random thrust, will maintain her swing to the same height i.e. constant amplitude. More energy will be fed to the swing if the father pushes in the same direction as the velocity. This will lead to greater input than loss in energy, producing increased amplitude of swing. In such a situation, although the father is not giving a hard push, the amplitude of swing can become very large—this phenomenon is called resonance. Once the push is stopped i.e. no energy input, the swing will come to a halt because of air resistance and friction—her father used to supply energy to compensate this loss.

If the girl, however, decides to swing faster and higher on her own, she has to pump her legs stretching forward then drawing them up underneath her. This would provide enough force to increase the height of swing's arc along with the enjoyment of a downward swing with increased velocity.

GPE and KE

Swings work with continuous inter conversion of gravitational potential energy (GPE) and kinetic energy (KE). GPE keeps track of work which one does against gravity in a swing. As the swing is pulled upward, it gains GPE because the earth tries to pull it back downward. Since GPE corresponds to the high part of the swing, the higher one goes on the swing, the more GPE is accumulated.

In contrast, KE is the fast part of the swing and relates to the speed for rushing back and forth. KE is stored as GPE at the top of each swing. Thus more height (i.e. more GPE) is attained with faster speed (i.e. more KE) of the swing.

On her own, the more energy she burns by pumping after pumping with her legs, her swing's GPE increases. As a result, an extra height will be added giving her a wider swing ride. Once she stops pumping, the swing slows down and eventually stops due to resistance from air friction.

Energy of the swing can thus be increased in two ways: 1) if someone pushes her swing which would increase KE of swing, or 2) on her own she can increase her GPE by pumping with her legs. By raising her legs at the top of each swing, she can raise the overall centre of mass her body, effectively raising the height of her swing.

Maths

Girl swinging like a pendulum

Pendulum rides are the normal swings which give the feeling of flying in a controlled manner. The dynamics of a swing is related to KE and GPE. At the top of a swing, KE is zero while GPE is maximum and is equal to m * g * h where m is the combined weight of the girl and swing seat, g is acceleration due to gravity and h is the vertical height above ground. As the mass swings down, it picks up as much KE as it loses the GPE. At the bottom of the swing, gravity accelerates the girl to her maximum speed (V), and all her GPE is converted to a maximum KE given by ½*m * V^2.

Based on conservation of energy, one can write, KE (top) + GPE (top) = KE (bottom) + GPE (bottom) i.e. $0 + m*g*h = \frac{1}{2}* m*V^2 + 0$, which would yield

$V = sqrt(2*g*h)$ i.e. $\sqrt{(2*g*h)}$. - - - (1)

Based on equation (1) and a knowledge of her vertical position from the ground as she swings, one can find out her speed at any point of the swing. The equation also shows independence of the speed on weight i.e. a small or a big girl will have just the same speed if released from the same height.

Girl going all the way around a swing i.e. 360 degree revolution

At the top of the swing, there is a tug of war between several forces. The rotational centripetal (inward) force on the girl must exceed pull by gravity so that the girl can be held up without falling down. Since the total vertical distance from the top of the swing to the ground is now 2*R, where R is the radius of the swing, the velocity at the top of the swing can be expressed by modifying equation (1) as

$V(top)^2 /R \geq g$

On travelling from the bottom to the top, KE continuously decreases with increasing GPE. The minimum speed required at the bottom of a circular trajectory for the girl to go all the way around the swing, can be calculated by relating KE at the bottom to the energy at the top of the swing: $\frac{1}{2}*m* V(bottom)^2 = \frac{1}{2}*m *V(top)^2 + m* g*2* R$, which would give rise to

$V(bottom) \geq \sqrt{(5*g*R)}$ - - - (2)

It is evident that the normal velocity at the bottom has to be raised in one impulse from $\sqrt{(2*g*R)}$ following equation (1) to more than double as per equation (2) so that the girl can go around the swing in full circle. In real practice, the present theoretical discussion still requires 100% support.

Science of Common Experiences

Bouncing Ball

Following article on "Girl on a Swing", basic knowledge behind the science of a bouncing ball is discussed.

Physics

Forces and energies involved

We all know that a rubber ball bouncing on a flat surface slowly comes to rest. The dynamics of a bouncing ball involves several natural forces such as gravity, mass and velocity, and two types of energy namely potential energy (related to position), and kinetic energy (related to motion). Each of these forces and energies has different part to play on the bouncing ball as it speeds towards a floor and then bounce back away from it.

As the ball falls through the air, the total energy of the system remains the same i.e. energy is neither gained nor lost (Law of Conservation of Energy). The potential energy of the ball at rest changes to kinetic energy when moving.

On colliding with the floor, the ball is deformed. For a perfectly elastic ball, it will quickly return to its original form and would bounce up from the floor (Newton's Third law of Motion: each action

28

has its equal and opposite reaction). However, the ball will not bounce back to its original height. This is because the total energy is not retained, but partly dissipated in the form of heat (via friction) and sound energy on hitting the floor. Typically, the maximum height decreases exponentially from bounce to bounce.

Assessment of the behaviour of a bouncing ball

A simplistic approach will be to assume that the ball always bounces up and down in a straight path, and its movement is only affected by acceleration due to gravity (g). Any resistance from air friction, or spinning from friction as it hits the ground are assumed to be negligible.

The downward speed of a free-falling ball will have a vertical component of velocity $Vy = 0$ since it is released from rest, while the horizontal component Vx will not change unless it bounces against a vertical surface like a wall. The value of Vy will be constantly changing due to gravity. If the ball hits the ground with a going down time of Td, its velocity will be given by

Vy (downward) = Initial Speed $+\frac{1}{2} * (g* Td^2)$

where the initial speed is of course zero. As soon as the ball hits the ground, its upward vertical speed will be given by a similar equation but with negative sign:

Vy (upward) = - V (bounce) $- \frac{1}{2} * (g* Tu^2)$

where V (bounce) is ball's velocity at the time of the bounce and Tu is the time of going upward before descending down again.

Maths

1). Total distance covered by Upward bounce only

When a ball bounces on a floor, it recovers a certain percentage of its original height, say 75% or 0.75 times the original height. This means that if a ball is dropped from an initial height of say 64 feet (for convenience of calculation), it would bounce up to reach a height of 0.75 * 64 i.e. 48 ft. On second hit to the floor, the ball will bounce up to 0.75 * 48 ft = 36 ft which is the same as 0.75 * 0.75 * 64 ft), and similarly the next vertical length would be 27ft and so on.

This process of obtaining successive maximum height by multiplying the previous height by the same value, is known in mathematics as geometric sequence. Thus the distance covered by a ball travelling upwards only are: 48, 36, 27 ft and so on. These are terms of an infinite geometric sequence with a general expression of as

a(n) / a (n-1) = 0.75

where n is the number of bounces with the nth term as "a(n)" being the length equal to 0.75 times the length before it.

One can compute the sum of any number of terms in an infinite geometric sequence as:

S = a* (1-r^n) / (1-r)

where, r is the ratio of the terms called common ratio or coefficient of restitution which is < 1, namely 0.75 here. Thus considering upward vertical movement only, sum of the distance covered by the bouncing ball after the third bounce would be

48 * (1 - 0.75 ^ 3))/ (1 - 0.75) = 111 ft.

For infinite number of bounces, n will be too big, resulting in r^n very small, making it essentially 0, leaving 1- r^n as 1. As a result, the sum of infinite number of bounces vertically upwards would be:

S∞ = a/ (1 - r)

which would be 48/(1-0.75) = 192ft. This means that if the ball bounces forever and ever, the total upward travelling length would never exceed 192 ft.

2). Combining total vertical distance by both upward and downward bounces

In order to calculate the total distance travelled, one has to consider the bounce twice the height because the ball drops the same distance as going up. Starting with 64ft as the length which the ball first travels downward only once, the rest of the distances namely 48, 36, 27 and so on are repeated twice. Thus each of these lengths is to be multiplied by 2 except for the first term when the ball travels only once. Thus the sum of total vertical distance travelled by the ball bouncing n number of times would be:

Sum = 64 + 2 X [48(1- 0.75 n)] / (1- 0.75) = 64 + 384 (1-0.75n).

Thus, for example, at the end of third bounce when n=3, the total vertical distance travelled by the bouncing ball would be 286 ft which is the same as

64+ [2 X 64 X 0.75)] + [2X 64 X 0.75 X 0.75] + [2 X 64 X 0.75X0.75 X 0.75)= 286ft.

Following the same argument as before, the maximum total vertical length the ball would travel after infinite number of bounces will be 64+ [2 X 192] = 448 ft which can never be exceeded in the present context.

3). Total time of bouncing

The time t taken by the ball for the first bounce is given by $t = \sqrt{(2 * a/g)}$ where a is the height of the first downward travel of the ball i.e. 64 ft and g is 32ft/sec^2. The time taken by the ball after it hits the floor for the first time would, therefore, be

$\sqrt{(2 * 64/ 32)} = 2sec.$

After hitting the floor for the first time, the ball rises to 48 ft which would take 1.7 sec, and similarly to reach subsequent heights of 36, 27ft, the time taken would be 1.5, 1.3 sec respectively for upward vertical movements only. The common factor here is $\sqrt{0.75} = 0.86$.

As before, the time taken by the ball to return to the floor after infinite bounces i.e. $n = \infty$ would be

$$2 + 2 \{1.7/(1 - 0.86)\}$$

which is about 26.3 sec—the time taken by the ball to travel only in upward direction. Therefore, considering both up and downwards travelling, the ball will take a total time of

$$2 + (2 \times 26.3) = 54.6 \text{ sec.}$$

This would imply that the ball after theoretically going through infinite bounces, would come to rest soon after 54 seconds. It would be interesting if the present theoretical consideration of geometric sequence leading to exponential model can be corroborated with data from practical experiment.

Science of Common Experiences

Shape of beverage can

Following articles on "Girl on a Swing" and "Bouncing Ball", basic science behind the shape of beverage (soft drink) can is discussed.

Physics

Curved (convex) walls

Consider a flat floppy paper. It has hardly any structural strength due to negligible moment of inertia around its bending axis. However, once the paper is rolled into a tube, curved surface is generated, and the rolled paper would now exert stiffness due to increased moment of inertia around the bending axis. It exemplifies a fundamental engineering principle that curved rather than flat walls are stronger due to high cross-sectional moment of inertia. This concept would apply to all sorts of containers including cans. The sides of a can are not flat, but round i.e. convex (bulging outward) for improved strength.

However, strength cannot be a prime issue, otherwise the manufacturers could go for stronger material such as steel instead of normally used aluminium. The reason for using aluminium alloy as the can material is because of its high strength/weight ratio, hence less weight and ultimately good money saving.

Concave bottom

The bottom of a pressurized beverage can is concave (hollowed inward) which helps to withstand a normal internal pressure of ~90 psi or 6.2 bar of the beverage. If flat, the internal pressure would tend to cause the bottom to bulge outward. In this deformed state, the cans would not be able stand upright. This is important because it would be impossible for distributors and grocers to stack a can upright if its bottom bulges out. Aluminium alloy with good mechanical strength is used for the base and body of cans. A typical alloy contains weight 1% Mg, 1%Mn,0.4%iron, 0.2% silicon and 0.15% copper with the rest in aluminium.

Incidentally, it is interesting to note that the bottom of a propane tank is concave which provides in-built safety. Any accidental build up of internal pressure leading to catastrophic failure can be avoided if the tank expands i.e. pops from concave to convex.

Flat top (lid)

Top of a beverage can has to be flat for convenience to get mouth around. The flat top similar to the concept of strength at the end of a fire hose holding back water, needs to be quite strong. This is achieved by reducing the amount of manganese (Mn), and increasing the magnesium (Mg) content of the aluminium alloy. Thus for the lid, Mg content is adjusted up to 2% while Mn is reduced to a trace level. This change in alloy content and use of thicker plate makes the lid stronger as well as heavier. The flat lid is also considerably thicker than the walls. To partially counterbalance the added weight, diameter of the lid is made less than the rest of the can body, producing a characteristic taper shape as is normal in cans today.

Optimising the shape of a can

It is established so far that a concave bottom with curved wall surface would be ideal shape of a can made in aluminum alloy. Mathematically speaking, an optimum shape can be achieved in the form of a sphere which has the smallest surface area. This shape would require least amount of material and hence lowest cost. But physical reality forbids use of any such shape for cans, because it is hard to distribute and stack up spherical shaped cans, and above all it is not easy to hold them to drink from.

A cut of 3-D (three dimensional) sphere into a 2-D slice will produce a circle. Such circle will be the best container shape in 2-D which in 3-D shapes up will provide the form of a cylinder. Can is thus nothing but a cylinder that can hold pressurized drink. The cylindrical shape enclosing required volume with least circumference for a given area would be the best shape for economy of materials for cans. Based on experience, the ratio of height to diameter ("aspect ratio") of a beverage can is normally preferred to be at least 2.0 to make it easy to drink from.

Maths

Designing a drinks can that will have the largest volume for a given surface area

In order to formulate a mathematical model, it is assumed that the beverage can has a cylindrical shape of height h and radius r. It has a top and bottom with each having a surface area S leading to a joint area of $2S = 2 * \pi(pi) * r^2$.

The circumference (C) of the can is given by $C = 2 * \pi * r$. The total surface area A of the can is a function of r and h i.e. A = g(r,h), and is equal to the joint surface area of the top and bottom plus the surface area (h*C) of the curved side:

A = g(r,h) = 2*S + h*C or h = (A - 2*S)/C.

The volume V is a function of r and h. It is a product of the height and the surface area of the top or bottom

V = V (r,h) = h*S = (A - 2 *S)*S /C.

It is obvious from the above equation that the volume of the can is largest when the circumference C is minimized for a fixed value of S. In order to determine the largest possible volume of the present cylindrical can with surface area A, we need to find out the maximum value of the function V(r,h) = h *S = h*π*r^2 which is subject to the constraint g(r,h) = A = 2*π*r^2 + 2* π*r*h

Using variational calculus (*Mathematical Methods for the Physical Sciences by R. Snieder, p.461,Cambridge University Press, 2004) and* the method of LaGrange multipliers *(Click in Google: LaGrangeMultipliers by S. Ellermeyer, June2,1998)*, the partial derivatives (∂) of the volume will be a LaGrange multiplier λ of the partial derivatives of the surface area, so that one needs to solve ∂V/∂r (r,h) = λ ∂g/∂r (r,h) which would give:

2*π *r*h = λ*(2*π*h + 4*π*r) - - - equation (1).

Similarly, ∂v/∂h(r,h) = λ ∂g/∂h (r,h) would give π*r^2 = λ*2*π*r - - - equation (2).

On simplification, equation (2) becomes λ = r/2 which when substituted into equation (1) would yield 2*r*h = r(h+2*r) or h= 2*r that would give a total surface area A as A = 2*π*r(2*r + r)

With r = √(A/6π) and since h= 2*r, one obtains h = 2√(A/6π). These values of r and h are the dimensions of a cylindrical drinks can (with top and bottom) whose surface area is A and whose volume is as large as possible.

A typical soft drink can has a diameter and height of 6.2 cm and 11.4 cm respectively with a corresponding volume (π*r^2*h) of 344 ml. The supplied volume of drink in the can is actually 330 ml—the difference of 14 ml must account for the concave shape of the base.

Science of Common Experiences

Soap Films & Bubbles

Following articles on "Girl on a Swing, Bouncing Ball and Shape of Beverage Can", basic science is discussed for soap films and bubbles.

Soap films and bubbles have fascinated humans with their beauty for a very long time. A wire frame pulled out of a soapy water, shows up a soap film stretching across the frame. Physically, this film being elastic in nature seeks smallest possible i.e. minimum surface area. The film reaches equilibrium once the surplus water drains away.

Besides enclosing zero volume, the mean curvature of a soap film must everywhere be equal to zero in equilibrium condition. The combination of minimal surface area with zero curvature prompted scientists including Leonardo da Vinci, Isaac Newton and Joseph Plateau to study soap films and bubbles with surprising arrays of discoveries in both physics and mathematics.

Difference between a soap film and a soap bubble

A soap film is surrounded by air and consists of two layers of soap molecules separated by a thin layer of fluid, such as water. Its thickness is of the order of nanometers. Besides being created at the contact surface of two soap bubbles, thin films can also be created by pulling out a metal frame dipped in a soap solution.

A bubble is a film, but a film is not a bubble. Soap film in a bubble wand once blown will form a hollow sphere of soap bubble with an iridescent surface. Bubbles can last for a few seconds before bursting either on their own or on contact with another object. Similar to the concept discussed earlier for beverage can, the spherical soap bubble tends to enclose a maximum volume possible for a given surface area.

Surface tension

The most important property in the present context is surface tension (γ) which can be defined as the strength with which a liquid pulls itself together. For water, a γ-value of 0.0756 N/m is too high to stretch it forming stable bubbles when shaken with air.

In comparison, a soapy liquid with a much lower γ-value (typically a third of plain water) makes it "weak" to pull itself together. As a result, once blown in, the air remains trapped longer inside a soap than a water bubble.

Physics

Why soap bubbles display colours?

The cause of iridescent colours of a soap bubble is due to interference of light which is similar to that observed in an oil slick on a wet ground. This phenomenon is, however, not the same as the splitting of light by a prism via refraction, or origin of colours in a rainbow which is caused by refraction of internally reflected light in air borne water droplets.

Initially, light is reflected from both the inner and outer surfaces of a soap bubble. The light rays that are reflected off the inner surface of the bubble travel further than those reflected off the outer surface. The wavelengths will interfere "constructively" or "destructively" depending on the distance travelled by the transmitted-and-reflected rays, the angle of the incident light and the thickness of the film. Soap bubbles thus develop colours via interference of white light which is made up of different colours corresponding to specific wavelengths.

Change of colour

When white light shines on a bubble, the film reflects a specific hue. The colours of a bubble developed due to interference of visible wavelengths, are dependent on the thickness of the soap film. A typical thickness of the walls of a soap bubble is about a micrometer, regardless of the diameter of the bubble.

It is exciting to realize that a child holding a bubble wand is playing with a soap film only of a few hundreds of nanonmeters thick. If the bubble wand is held vertical, a beautiful band of colour appears. This is because in the vertical position of the wand, the film is thicker near the bottom and thinner near the top as the liquid in the film gravitates downward. Blue-green colours dominate in thicker films and yellow hues in thinner films. A bubble can also get progressively thinner as it dries out due to evaporation before finally bursting.

Maths

A study of the problems of predicting soap film geometry has led to multiple branches of advanced mathematics such as analogue solutions to minimization problems and development of calculus of variations.

Basic consideration

Consider an inflated rubber balloon. The inside air pressure is greater than on the outside in order to counterbalance the contraction force of the stretched rubber. On a similar note, when a soap bubble is creared, its stability is maintained by having greater inside than outside pressure.

However, the inside pressure cannot go on increasing with time, which would, otherwise, lead to bursting of the bubble. There is thus another pressure operating namely the γ-value of the soap solution that pulls the bubble towards its centre, making it to contract to as small as possible. Thus for the stability of a soap bubble, a balance is maintained between the expansive pressure from the inside and the contractive pressures of air from the outside along with γ-value of the soap solution. The radius R of the spherical bubble also influences its stability.

Force inside a soap bubble

Consider an imaginary cross-section through the centre of the bubble. Along the cross-section, ignoring the very slight difference in inner and outer radius, the circular cross-sectional area of the hemisphere is $\pi*R^2$. Inside the bubble, a pressure P is acting over the entire cross-section area, resulting in a total force of $P*\pi*R^2$.

Force due to surface tension

The circumference of the cross-section of the bubble is given by $2*\pi*R$. Each inner and outer surface will have a contractive pressure controlled by γ along the entire length. Thus the magnitude of the force due to each surface of the bubble is the product of γ and the circumference of the circular edge i.e. $\gamma*(2*\pi*R)$. The total force from both inner and outer surface is twice this amount i.e. $2*\gamma(2*\pi*R)$.

Combined forces

For a stable bubble, the two forces namely the expansive inside air pressure and the contractive effect due to surface tension must be equal and opposite. Following Newton's second law of motion, sum of these forces must be zero. In other words,

$p*\pi*R^2 = 2*\gamma (2*\pi*R)$.

Solving the equation for the pressure inside the bubble, one obtains $P = 4*\gamma/R$. Obviously, this is a simplified analysis where the pressure outside the bubble (Po) is assumed to be zero. However, this result can easily be extended to obtain the pressure difference between the inside and outside of the bubble:

$P - P_0 = 4*\gamma / R$ - -- - (Spherical soap bubble).

This result implies that the difference in pressure depends on the surface tension and the radius of the bubble. Also, it is noticed that a greater pressure exists inside a smaller soap bubble (smaller value of R) than inside a larger one.

In contrast, for a bubble of a liquid such as water droplet, there is no access for air. As a result, one has to consider, instead of two, only one surface namely the outside one for a spherical water droplet. Consequently, the above equation for a soap bubble has to be modified as

$P - P_0 = 2*\gamma / R$ ---- (Spherical water droplet)

which is generally known as Young—Laplace equation for a spherical liquid drop.

It is hoped that the present text has managed to bring together a mixture of simple physics with basic mathematics that underlines the interesting subject of soap films and bubbles.

Science of Common Experiences

Lift of Winged Flight

Following articles on"Girl on a Swing, Bouncing Ball,Shape of Beverage Can and Soap Film and Bubble", Lift of Winged Flight is the last topic in the series.

Flight is a phenomenon that has long been a part of the natural world, namely the birds gliding with their wings outstretched. Analytically, flights of birds and man-made aircraft rely basically on the principles of physical science.

Physics

Four basic physical forces

The flight of an aircraft is made possible by a careful balance of four physical forces namely lift against weight, and drag against thrust. Lift manifests itself as a mechanical (aerodynamic) force generated by a solid object (aircraft or bird) moving through a fluid (air), which directly opposes downward gravitational force of weight.

For an aircraft, wings are used for lift, the propellers and jet engines provide the thrust while drag caused by moving through air is reduced by plane's smooth shape. Its weight is controlled by choosing suitable construction materials (high strength/weight ratio). A high lift-to-drag ratio of 1or 2 orders of magnitude would propel the wings through the air at a sufficient height.

Physical description of lift mechanism

There are several mechanisms of lift of aircraft by airfoils i.e. airplane wings. Two most popular ones are described that include Bernoulli's equation based on conservation of energy in fluids (air), and complex combination of Newton's laws and conservation of momentum. The attack angle in air via tilted wing (Coanda effect) is also an important contributory factor.

Bernoulli's explanation of lift

A combination of Bernoulli and the continuity equations provides an intuitive guide for analyzing fluid (air) flows. Accordingly, when air passes over a solid body such as airfoil, the streamlines get closer together with the flow velocity increasing. Following Bernoulli's principle, a fast flowing air, for example over the top rather than bottom of a airfoil/wing, would decrease the surrounding air pressure.

This creation of differential air pressure between the top and bottom of a wing helps to lift an aircraft. However, in order to explain why the air goes faster over the top of the wing, one has to resort to the geometric argument that the distance the air must travel is directly related to its speed. A simple estimation shows the distance over the top of a wing to be about 50% longer than under the bottom.

It is difficult to appreciate how the above principle of low-pressure, high-velocity air on the top side and high-pressure, low-velocity air on the bottom side of the wing could explain "inverted" flight. The truth is that the top-side air travels significantly faster, hence creating lower pressure than the bottom-side air, and they never recombine. Also, the angle of attack of air on the airplane wings has a profound effect.

Newton's laws and lift

Basic Newtonian principles of aerodynamic lift and propulsion are considered. Newton's first law states that velocity of an object changes only when its mass is acted upon by an applied force. That means, if one sees a bend in the flow of air, or if air originally at rest is accelerated into motion, there is a force acting on it. Forward propulsion, of propellers and jets, is gained by acceleration of air mass at the rear.

According to Newton's third law, every action has an equal and opposite reaction. For an aircraft, what the wings do to the air is the action while the lift is the reaction. A normal atmospheric mass distribution is maintained via upward recirculation of air mass at a rate equal to the rate at which it is displaced downward in the lift process. The wing must change something of the air for the lift. This involves changes in air's momentum, resulting in forces on the wing by diverting down lots of air.

Following an alternative form of Newton's secod law which relates force (F) on an object as a product of its mass (m) and acceleration (a) i.e. $F = m* a$, the lift of a wing can be stated to be proportional to the amount of air diverted downwards multiplied by the downward velocity of the air.

Maths

Mathematical aerodynamics used by aeronautical engineers, involves complex mathematics and/or computer simulations to calculate lift of a wing. In the present context, a simple approach is discussed by considering an important parameter called lift coeffeicient.

Lift coefficient

The effectiveness of lift of an aircraft depends on a number of basic parameters such as density (ρ), velocity (V) and surface area of the wing (S). Mathematically, the modern equation for lift is expressed as

Lift (P) = CL * (½*ρ *V^2)*S

where CL is a number denoting lift coefficient. The quantity in parentheses is proportional to dynamic pressure. In designing an aircraft wing, it is usually advantageous to get the lift coefficient as high as possible. The CL term determined experimentally, is used by aerodynamicists to model all of the complex dependencies on the lift of an aircraft. These include considerations of shape of the aircraft body, its inclination to the airflow, and the air flow conditions such as viscosity and compressibility of air.

Common expectation is borne out by the above equation. For example, a larger lift can be expected from a larger airflow velocity as well as from heavier air when the airflow deflected by the wing has a larger momentum. For airports built in high altitude will have relatively rarefied lighter air especially in the summer. Since the surface area of aircraft wings is fixed by design, longer runway is needed for such airports so that an aircraft can lift after acquiring acceleration necessary for larger take-off velocity.

Dimensional analysis

Approach via dimensional analysis can explain the relationships between different physical parameters that are important in the design of scale experiments such as a model of an aircraft in a wind tunnel. Eventually, all parameters need to be scaled up appropriately in accordance with the size of the aircraft so that the physics remains unaltered by scaling.

Based on Buckinham π(pi) theorem, dimensionless numbers can be shown to be the same for a scaled model as for a real aircraft. Because lift (denoted by P) is a force, it has a dimension. Following Newton's second law i.e. force,F = m*a, one can write dimensionally $P \sim [F\} = [ML/T2]$ where M, L and T are the fundamental quantities of mass, length and time respectively. The variable parameters of lift namely ρ, V and S can be written as $\rho = [M/L3]$, $V = L/T$ and $S = L2$. Using these dimensions, the lift coefficient would turn out to be a dimensionless number.

Final Comments on the Present Series

A physical theory is a model of everyday events. Its quality is judged by its ability to make new predictions which can be verified by new observations. A physical theory differs from a mathematical theorem in that while both are based on some form of axioms, judgement of mathematical applicability is, however, not based on agreement with any experimental results. Standard scientific and mathematical practice seems to invoke mathematical explanations.

Science of Natural Phenomena

Colourful Optics of Nature

Following the series on Science of Common Experiences, some of the natural phenomena are explained using basic physics and mathematics.

The laws of physics and mathematics (maths) complement each other in explaining the orders in nature. Any explanation of natural phenomena is based on two basic issues:

1). collection of data by physicists from experiment and observation, and
2). inference of results by mathematical reasoning.

Statements from maths have to be logically true, while predictions of physics must match observed and experimental results. Historically, physics and maths are inextricably linked with maths as a language and a source of insight into physical sciences.

Following similar trend as in the previous series on Science of Common Experiences several interesting natural phenomena are explained in the present series using basic physics and maths.

Colourful Optics of Nature

Nature has been perfecting the design of optical systems for 500 million years since the evolutionary "big bang" took place, followed by rapid diversion of animal life and plants. Manipulation of light started to influence the survival of biological systems such as aquatic creatures, insects, birds and flora that used submicron structures to produce striking optical effects.

Physics of optics

Optics is the branch of physics that deals with the behaviour and properties of light.

Colour Systems of Nature

Display of multitude of colours has always captured the minds of humans from Aristotle (360 BC) to contemporary scientists. Pigments are responsible for most colours of birds' feathers (avian plumage). However, blues seen in *macaws* and snowy white feathers of Bali *mynahs* are structural colours which result from light reflective properties of feathers. Another form of colour is the rich hues of iridescence which is similar to structural colour except that hues are altered with the angle of observation.

Pigmentation

Pigments are molecules that differentially absorb and emit wavelengths of visible light. Pigmentary colours are common in feather, egg shells, dermis and eyes of birds.

Structural colour

Structural colour results from interaction of light with physical structures that have variations on a spatial scale comparable with the wavelength of light. Robert Hooke (1665) and Isaac Newton (1730) initiated the work on animal optics by correctly reporting that colours of insects (silverfish scales) and peacock feathers were due to their physical structure rather than pigmentation.

The bright wing surfaces of butterflies, feather barbules and barbs of birds, peacock's tail feathers, fish scales soap bubbles and oils slicks are all examples of brilliant colours that have a structural origin. The same feature applies to the interaction of light with nanometre-scaled biological structures that would generate structural colour through interference property of (reflected or diffracted) light.

Structural vs. Pigmentary Colours

Structural colours, unlike pigments, are produced by physical interaction of light with structural layers. Turacos alone among all bird families can produce their own special green pigment. Other birds rely on a combination of both structural and pigment colours for their green feathers created by a combination of a structural blue hue and a superficial yellow carotenoid pigment (blue + yellow = green).

A red or yellow pigmented feather will retain its original colour when ground into a powder, because pigments are not damaged even though the feather structure is destroyed. In a similar condition, structural colour of a blue feather becomes, however, dark since such colour depends entirely upon reflective properly.

Iridescence

Originating from the Greek word *iris,* meaning rainbow, iridescence is an optical surface phenomenon characterised by dynamic, flashy spectrum of colours with green, red and orange as the most common varieties. Soap bubbles and oil slicks show iridescence. Butterflies are renowned for their brightly iridescent wing surfaces whose nanostructures interfere with incident light producing variety of visual colouring effects.

In iridescence, change of hue occurs with the type of sample used, the quality of lighting and most of all the viewing angle. Also, smooth and wet surfaces are more conducive to iridescent oil slick compared to rough and dry surfaces that tend to limit iridescence.

Structural colour vs. Iridescence

Iridescence is produced not by pigmentation but by interference of light resulting in multiple reflection of light within the bulk of the body (more about it later on). However, association with physical structure of a material makes iridescence sometimes referred to as structural colour which, of course, relies on the relation between crystal lattice spacing of the object and wavelength of light. In short, one can say that iridescence requires structural colour, but the converse is not necessarily true.

Unlike structural colour, the observed change of iridescent hue varies with the angle from which a surface is viewed, and can occur over a range of wavelengths. Generally, iridescent feathers appear very bright and colourful compared with a structural feather viewed under the same light. Structural blue feathers remain blue to the observer even when the feather is rotated in relation to the light source. In comparison, colouring of iridescent feathers varies and can become black with the angle of rotation of light.

Mechanics of natural colouring

Visible light (photon) is composed of many colours with wavelengths (λ) ranging from the longest ~ 700 nm for red to the shortest ~ 400 nm for blue/violet light. Blue light with very short λ, is reflected easily than other colours. (The familiar "blue sky" phenomenon results from preferential scattering of blue light by miniscule particles in earth's atmosphere.)

The colours of visible spectrum are reflected while encountering particles of same or larger diameter than their own λ—values. For example, particles ~ 400 nm diameter will selectively reflect blue light photons, while allowing other light photons to pass by. The reflected light photons provide the perceived image of colour to an observer's eyes. Blue colours in the plumage of most bird species results from preferential scattering of blue light by feather structure. In contrast, white plumage of the Bali *mynah*, is produced when all λ–values of visible light are reflected by the feather—a typical example of the reflective property of structural colour.

Interference of light in natural films (soap bubble and oil slick)

Thin film optics with thickness of the order of average λ of visible light ($\sim$ 500 nm) provides remarkable reflective property. This is due to interference effect of light wave, and the difference in refractive indices between the film and air. As a result, iridescent rainbow colours are seen in thin film of soap bubbles or oil slicks on a wet road.

For soap bubble as an example, some of the incident light bounce off the outer surface while the non-reflected light enters the bubble and re-emerges after being reflected back and forth between the top and bottom surfaces. Although the reflected light waves from either bottom or top surfaces travel in the same direction, there is, of course, a difference in distance covered.

Consequently, the reflected light wave from the bottom surface may be in-phase (strong reflection due to constructive reinforcement of light wave) or out of phase (weak reflection due to destructive cancellation) with respect to the light wave reflected from the upper surface. Thus when white light with full spectrum of colours, impinges on a soap bubble, it will be reflected with a hue that changes with thickness due to in-phase or out of phase of wavelengths or colours.

Photonics and nanostructure in nature

The word "photonics" is derived from the Greek word "photos" meaning light, and is related to the properties and transmission of photons i.e. light. On a finer scale, nanophotonics is a recently emerged field of nanotechnology covering aquatic systems, insects, birds and flora. It considers the behaviour of light on a nanometre scale controlling the interactions of light with matter on λ and sub-λ scales.

Photonic crystals and natural colour

Identification of "photonic crystals" in creature became known only about three decades ago. It is now established that photonic crystals are used by light-sensitive *Ophiocoma wendtii* species of brittle star to collect light, *Morpho* butterflies to exhibit their striking blue wings, and peacocks to display brilliant, diverse colours of blue, green, yellow and brown.

By definition, photonic crystals represent periodic optical nanostructures that are designed to affect the propagation of light. For photonic crystals to operate in the visible part of the spectrum, the periodicity of their structures has to be of the same length-scale as half the wavelength of light waves i.e. ~200nm (blue) to 350 nm (red) since the physical aspect of optical phenomena relies basically on diffraction property.

There is a well-defined frequency range (called photonic band gaps, PBG) in photonic crystals which does not allow light to propagate through the structure but reflects it back. Natural systems employ remarkable three dimensional photonic crystals which reflect bright colour via natural PBG systems.

Maths

A simple mathematical model of iridescence is presented. It is based on reflection and transmission of light waves through a combined stack of thin reflective layers such as the wings of butterflies and peacock's tail feather.

Reflection coefficient (R) of the combined stack is defined as the ratio of the strengths of the reflected and incident waves, and the corresponding transmission coefficient (T) as the ratio of the strengths of transmission and incident waves. The overall values of both R and T can be derived from the corresponding coefficients of individual thin layer using Born approximation.

Mathematically, the light waves propagating through a stack of layers are expected to be reflected repeatedly i.e. bouncing back and forth between layers. Occurrence of multiple reflections affects the calculations of both R and T of the combined stack. This would imply that R and T cannot be derived by simple additions of similar properties of individual layers.

An evaluation of R is complicated by the predominant effect of the top layer compared to the layers in the bulk on reflectivity. However, T can be shown to be the product of transmission coefficients of individual layers provided the product of the reflection coefficients of the layers is less than 1.

Any discussion on the mathematical treatment of diffraction and interference phenomena is far too complex to be of interest in the present context, and therefore not considered.

Final comments

History suggests that there is no more experienced engineer than nature itself when it comes to highly advanced optical systems. Physical optics dealing with interference, refractive index and diffraction of light can explain the visual effects of natural colours. Present review illustrates how light is manipulated by living creatures by exploiting photonic structures.

Science of Natural Phenomena

Natural Vortex

Following article on "Colourful Optics of Nature" in the present series on natural phenomena, basic science is discussed for natural vortices ranging from bathtub drainage to cyclone and tornado.

Vortex is a generic term for swirling movement. It describes various circular forces that exist in the universe. The drainage of water from a bathtub, kitchen sink or industrial reservoir creates a natural phenomenon of hydraulic vortex (pl. vortices) in the drain.

Vortex similar to the bathtub situation is often encountered in nature on a much larger scale such as atmospheric vortices in cyclone/ hurricane/typhoon and tornado.

Physics

The basic science of vortex is discussed for water flow in bathtubs and air circulation in cyclones and tornadoes.

Primary factors influencing vortex movement
Earth's rotation

The ground in the northern hemisphere rotates counter clockwise as the Earth spins eastward beneath our feet. Everything around us including any fluid (air or water) gyrates at the same time aided by angular momentum. The daily rotation of Earth causes vortices, specially the atmospheric types, to spiral anti-clockwise in the northern hemisphere and clockwise in the southern hemisphere. Does this mean that a storm would have its rotation changed if it manages to cross over the equator? Perhaps, the air currents will not allow the storm to cross over equatorial lines.

Coriolis force

If the Earth did not rotate, air would simply move across the globe blowing straight from the tropics to the poles, or vice versa. But this does not happen. Winds and storms follow curved paths because of the Earth's daily spin on its invisible north-south axis. This phenomenon is called the Coriolis effect.

If we consider an object on the equator completing a circuit around the Earth in one day, it would have to travel Earth's circumference at an average speed of ~1000 miles/h. If the object starts from a point longitudinally away from the equator, both the speed and distance covered would decrease and eventually become negligible at the pole.

The development of differential speed by moving away from the equator would affect movement of a fluid that would be deflected from its path as seen by an observer on Earth. Such deflection of the inertia (power of resistance to change of motion) of fluid due to Earth's rotation is called the Coriolis force which is quite small because of Earth's very slow one complete rotation per day.

The Coriolis force despite being weak is noticeable for largish events such as atmospheric vortices. However, any deflection of fluid disappears at the equator where the Coriolis force due to Earth's rotation is negligible.

Rosby number (Ro)

Rosby number is the ratio of inertia to Coriolis force. It determines the forces at play in a given situation whether it is water spinning down a drain or air swirling in a cyclone. A large Ro (>> 1) signifies a system where inertial and centrifugal forces dominate as in tornadoes (Ro~10^3). However, for Ro << 1, motion of a fluid is affected by Coriolis force i.e. rotation of the system.

Specific cases

Vortex in bathtub

Bathtub vortex is characterized by intense axial down-flow. When the cross-section area of the drain hole relative to the bathtub is sufficiently small, the angular momentum dominates. The speed and rate of rotation of water increase as it approaches the hole centre, but decrease progressively with distance from the centre.

At first, the water particles running towards the drain are pushed off to one side because of momentum gathered by other water particles rushing toward the drain at the same time. This deflection along with the principle of conservation of angular momentum sets a chain reaction with water particles spiralling down the plughole. The strong rotating effect ultimately creates a vortex.

For low Ro (~0.1) and bathtub vortex length of 0.1 meters, the Coriolis force is estimated to be very small (< 10^-6). Being very weak, the Coriolis force can affect the direction of vortex only if the motion of water in the bathtub is less than Earth's daily rotation of 0.00001 per second. This apparently unrealistic ideal condition has also to be supplemented by elimination of thermal current, vibration and any other external disturbances.

Vortex in Cyclone/Hurricane/Typhoon and Tornado

Terminology

Cyclones, hurricanes and typhoons are all similar weather phenomena of violent storms. They are called differently in different parts of the world. In the South Pacific and Indian oceans including the Bay of Bengal, the term "cyclone" is used (affecting Australia and Bangladesh for examples). The same type of climatic disturbance in the Atlantic and Northeast Pacific is called "hurricane" (occurring in the U.S.), or "typhoon" in the Northwest Pacific (threatening Japan and Philippines).

In contrast, tornado is a violently rotating column of air in contact with the surface. It is often visible as a funnel cloud. For a vortex to be classified as tornado, it must be literally in contact with the ground and the cloud base.

Vortex in cyclone

Cyclones, as the deepest of all low-pressure weather systems, are developed over warm tropical ocean near the equator. Air heated by the sun rises swiftly creating low pressure voids. Cool air rushing in to fill the voids bends inwards because of Earth's constant turning on its axis (Coriolis force). The inward movement of the wind spirals upwardly faster and faster forming a huge circle that would lead to atmospheric vortex. Water vapour and droplets are sucked up from the ocean surface into the vortex which ends up with devastating effect once enough power is acquired.

Air flowing around a cyclone spins counter-clockwise in the northern hemisphere and clockwise in southern hemisphere (as does the Earth itself). If the Earth did not rotate, the air would flow directly towards the low pressure centre where a calm, cloudless atmosphere called the "eye" exists with light winds and no rain.

Because the Coriolis force initiates and maintains cyclone rotation, such storm could rarely form or move near the equator within ~ 5 degrees where the Coriolis effect is weakest. Cyclones thus chiefly occur in the mid-latitude belts of both hemispheres.

Vortex in tornado

Tornadoes begin at low-pressure areas along Earth's surface that would draw in cooler high-pressure air around them. Once brought in, the cooler air pushes the low-pressure air to higher altitude where it gets hotter and is still forced upwards by the air behind it. This results in the cooler surrounding air, at pressure of as much as 10 % higher than inside a tornado, to rush into creating a tornado even faster, causing the air to rotate even more. Consequently, a tornado picks up speed forming cylindrical vortex. A classic example of tornado development is the collision between the warm moist air from the Gulf of Mexico and the cooler, drier air from the northern Plains and the Rockies.

Tornadoes apparently contain enormous amount of energy and power around 500 million horsepower, mostly as kinetic energy of rotation of air moving upward inside a tornado. They look like giant vortex-funnels that actually descend from the cloud above them, rather than the optical illusion of starting from the ground.

Maths of vortex formation

Vector analysis

It is difficult to convince people of the rotation of bathtub vortex. One mathematical approach would be to consider the equations of fluid flow on a rotating Earth and compute numerically by feeding in all details of flow in a bathtub. This kind of numerical simulation lacks physical insight.

A vector analysis representing magnitude and direction of forces can provide a simplistic way of estimating the relative strengths of forces that act on a fluid forming vortex. Let us denote the rotational velocity of water in the bathtub and Earth's rotating velocity as ω and Ω respectively. The primary forces forming a vortex would include pressure force (Fx) due to the pressure gradient in the fluid, the Coriolis force (Fy) from Earth's rotation, and the centrifugal force (Fz) due to the circular motion of the fluid in the vortex.

Both Fx and Fz, unlike Fy, are independent of the position of vortex on the Earth. They will point inward (Fx) and outward (Fz) in relation to the vortex. This means that any asymmetry between the behaviour of a vortex in the two hemispheres must be due to the Coriolis effect (Fy).

Based on vector analysis, both Fy and Fz, unlike Fx pointing toward the vortex, would always direct away (outward) from the vortex. Fx needs to be balanced by Fy or Fz or by both to restore balance of forces,

The ratio of the strengths of the Coriolis (Fy) and centrifugal (Fz) forces can be shown to be of the order of the ratio of the rotation rate of the Earth to the rotation rate of the vortex in the bathtub i.e. Ω/ω. Since the Earth rotates once a day and if it is assumed that the vortex rotates once per second, Ω/ω would be 0.00001. This implies that the Coriolis force is much smaller than other forces operating on the vortex. It also follows from Fy/Fz ~ Ω/ω that both these forces would be of the same order of magnitude when the rotation rate of water in the bathtub is slow enough to be equal to Earth's rotation.

Advection-diffusion equation (ADE)

ADE describes a broad range of natural phenomena, and will be mentioned here very briefly because of its highly complex mathematical nature. ADE depending on application can also be called convection-diffusion equation. It describes mixing of water within a vortex that would involve convection (advection) and diffusion of water. Advection by circulating flow generates spirals of water which are so tightly wound that molecular diffusion becomes an important issue. The interaction between advection and diffusion is subtle and can only be understood through mathematical and computational analysis of ADE.

Final comments

Vortices in all different scales and intensities are ubiquitous in the Earth's atmosphere. Although it is a common phenomenon, its structure, formation and dynamics are still not completely understood despite years of intense fascination. Vortices generated in both cyclones and tornadoes are based on a similar principle as bathtub vortex, except air instead of water being the fluid, and, of course, occurring with a much larger motion of the fluid.

Science of Natural Phenomena

Lightning

Following articles on Colourful Optics of Nature and Natural Vortex, basic science of lightning is discussed as the third article in the series on natural phenomena.

Lightning is one of the oldest observed natural phenomena on Earth. Early humans considered lightning as a weapon of their gods. It was the Norse god Thor, the Greek god Zeus, and the Roman god Jupiter who wielded the mighty lightning flash to keep man in his place.

Based on his famous kite flight experiment in a stormy weather, Benjamin Franklin in 1752 was first to prove the electrical nature of lightning. Despite some perpetual debate in the scientific community about how the electrification of clouds actually occurs, part of the mystery started to unfold in the last two centuries. The present article discusses the basic physics and mathematics of lightning.

Basic considerations

Nature of lightning

An electrostatic type of electricity is developed in lightning, which is similar to that created by walking on a synthetic carpet on a dry day and then touching a conducting surface like metal. The scuffling of feet causes the human body to pick up electrons with negative (-)ve charges. On touching a metal characterized by positive (+)ve charges, the electrons jump across the small gap between the body and the metal producing a tiny electrostatic electricity shock which may end up as a spark for big enough discharges.

For lightning discharge, a cloud intially needs to acquire (-)ve charge by say frictional rubbing between falling raindrops and the air. The charged up cloud next seek (+)ve charges available from places like the ground on the Earth. Lightning develops in the form of a giant electrostatic spark when the gap between the two opposite charges closes down.

Lightning can occur during thunderstorm, volcanic eruption or dust storm. The precursor of lightning is the development of (+)ve and (-) ve charges within a storm cloud.

Properties of lightning

Each spark of lightning strike/bolt can extend over several tens of kilometres, soar to temperatures over 50,000 degrees Fahrenheit in a few milliseconds, carry an electric potential of more than 100 million volts and a current as strong as 100,000 amps. In comparison. a typical household needs around 240 volts and 220 amps.

Lightning and thunder

Introduction

Lightning represents a great deal of energy, and where there is lightning, there is thunder. The whole concept conforms to the law of conservation of energy of not creating or destroying energy but transforming, in this case, the electrical energy in to light (lightning) and sound (thunder).

Lightning is what we see, and thunder is what we hear, but both occur because of electrical discharge in the atmosphere. Lightning is seen first before hearing the thunder, because light travels (~300,00km/sec) much faster than sound (~0.34km/sec).

Sequence of events

The latest idea about the lightning-thunder phenomena is based on the steam explosion theory developed in the second half of the 19th century. The high temperature generated by lightning heats up the air too quickly (in a few milliseconds). This results in a sudden and violent expansion of super-heated air at a rate faster than the speed of sound, similar to a sonic boom. The shock wave thus produced extends outward to eventually reach an ordinary audible sound wave of thunder.

The intensity and type of sound (sharp or rumblig) depends upon atmospheric conditions and distance between lightning and the listener. The closer the lightning, the louder the thunder.

Physics

Ionization of clouds

There are two hypotheses of ionization of the storm clouds assisted by turbulent wind environment of a thunderstorm.

In the friction charging mechanism, the suspended water droplets and ice particles in stormy clouds move and whirl about in a turbulent fashion. Any rising of ground water via evaporation forms additional droplets of water clusters which collide with those already present within the clouds and rip off their electrons. This results in, like frictional charging, a separation of (-)ve electrons and positively charged water droplets.

A more detailed concept is expressed in the freezing hypothesis. Accordingly, the rising moisture from ground is frozen in colder temperatures at higher altitudes. The frozen particles cluster tightly together to become negatively charged while the outer droplets acquire (+)ve charges. Air currents within the clouds rip the outer portions of the clusters and carry them upward with (+)ve charges toward the top of the clouds (updrafts). The frozen portion of the droplets with their (-)ve charges tends to gravitate down towards the bottom of the storm cloud (downdrafts). These actions result in ionizing the the storm clouds.

Mechanics of lightning

Updrafts

The lightning flashes originating from upper regions of cloud are (+)ve. They carryhigher charge and last longer than (-)ve downdraft strikes, causing more damage to power and electricity infrastructure, and also starting more forest fires than (-)ve strikes.

Downdrafts

The vast majority (about 95%) of downdraft lightning are (-)ve with a net transfer of (-)ve charges from the cloud to the ground. The negatively charged cloud induces opposite i.e. (+)ve charges on tall objects on the ground such as trees, telephonepoles and buildings. As the electrostatic charge in a storm cloud gradually builds up, the electric field surrounding a cloud becomes stronger.

Normally, the air surrounding a cloud acts as an insulator to prevent any discharge of electrons to the Earth. But as the electric field gets stronger, the surrounding air molecules become ionized by shredding off their electrons from the outer shells. This makes the the air molecules behaving like conductive plasmons by turning into a soup of (+)ve ions and free electrons.

Since the potential at the bottom of the cloud is much lower than that on the ground, the (-)ve charges from the cloud channel their way through the air plasmons towards the positively charged ground in a series of steps (stepped leader). As a result, streams of (+)ve charges from the ground start to move upward (streamer or upward leader) along tall objects.

When stepped leader and streamer of opposite charges meet, the discharge releases an enormous amount of energy in the form of lightning or flashes known as the return strokes. They are called return strokes because the flash originates in the cloud, not at the ground. Such a stroke is much larger and brighter and more damaging than the stepped leader, and shoots up almost instantaneously to the cloud following the path of the stepped leader.

It is the return stroke that we see as lightning with bright luminosity in the sky which travels about 60,000 miles per second back towards the cloud. The whole process occurs so quickly (in less than a second) that the lightning appears to travel from cloud to ground. In fact, the opposite is true i.e. lightning occurs due to return strokes travelling from ground to the cloud.

Types of lightning

The ionization of storm cloud basically results in creating and growing electric fields between the cloud and the ground (C-G), and within the cloud itself i.e. cloud-to-cloud (C-C). Flashes of lightning between a thunderstorm and earth are C-G type, while lightning occurring as a bright flicker within a thunderstorm is classed as C-C type. There is roughly 5 to 10 times more C-C than C-G flashes.

Summing up

Lightning is the transfer of significant charge between two charged objects. Among the various classes of lightning discharges, the two basic types are C-C and C-G types. A C-G flash is typically composed of a sequence of individual C-G return strikes.

Protection from lightning

Besides proving electrical property of lightning, Benjamin Franklin also considered protection of buildings and other structures from damages by lightning. He came up with the idea of installing an elevated metallic rod next to a building, but earthed to the ground. In this way, any damage due to lightning can be prevented by conducting away the electrostatic charge straight into the ground without passing through the building. Since then some 200 years ago, the concept of lightning rod is still widely used today at air terminals as lightning protection systems.

Based on models of lightning strikes as functions of strike distance (classic model) and height of object (Eriksson model), analytical tools are developed to protect pylons and air transmission lines against damages from lightning.

Maths

The physical characteristics of lightning are very complicated with many controversies and ambiguities. Various mathematical models have been suggested to explain the different structural forms of lightning, such as graph theory for tree structures, dynamical systems for bifurcation type, or complicated differential equations for the inherent physics involved.

The paths of lightning, like many of the shapes in nature, are however, fractal i.e. shaped inirregular, chaotictrajectories instead of perfect lines and curves. This fractal nature is similar to the Laplace growth of the phenomenon. Based on scaler and vector analyses, one can show from computing Laplace equation that lightning must start at the charges that generate the field or at the Earth's surface, rather than starting in mid-air far away from Earth's surface. Meteorologists use a combination of physics and maths to track when and where lightning will strike.

Final comments

Lightning causes vast amount of damages to humans and materials. Despite its crucial importance, lightning is still poorly understood. This is because it involves a wide range of physics such as electromagnetism, atomic physics, plasma physics, meteorology etc. Not only the theory is complex, the observable phenomena range in scale from as low as sub-millimeter electron avalanches to giant ~100 km lightning flashes.

Science of Natural Phenomena

Earthquake

Following articles on Colourful Optics of Nature, Natural Vortex and Lightning, basic physics and maths of earthquakes are discussed in this fourth article of the series on natural phenomena.

Most earthquakes tragically demonstrate the suddenness with which they occur and the devastations they cause. Earthquakes occur because of sudden movement of Earth's surface. A brief knowledge of Earth's structure is outlined first for an eventual appreciation of the mechanics of earthquakes.

Earth's structure

In the early part of the 20th century, geologists studied the earthquake-generated vibrations (seismic waves) to learn the structure of Earth's interior which basically consists of several distinct layers: crust, lithosphere, asthenosphere, mantle and core.

Crust

Crust is Earth's outermost, thinnest, hard, rigid layer of only a few miles thick under the oceans and averaging 20 miles thick under the continents.

Lithosphere/Tectonic plate

Beneath the crust lies lithosphere or tectonic plate which covers the entire surface of the Earth from the top of Mount Everest to the bottom of the deepest part of the world's oceans (Mariana Trench). It comprises of some crust and uppermost part of the mantle.

Asthenosphere

This is a hot, malleable, semi liquid zone (~ 120 miles thick) over which the plates of lithosphere move or float.

Mantle

Mantle is a vast ocean of hot, dense, semisolid (pasty) rock, and forms the largest part of earth's volume. It is subdivided into upper and lower regions and extends down to a depth of about 1800 miles. Thus although the ground under our feet appear solid, we are actually standing on a relatively thin crust of rock below which is the mantle. Although the mantle is largely hidden from our view, it becomes visible in places where crack opens up, allowing the molten rock to escape in the form of volcanoes. The liquid rock pouring out of a volcano is the same as in the mantle.

Mantle convection

When the mantle rocks near the radioactive core are heated, they become less dense than the cooler, uppermost mantle. As a result, the warmer rocks rise while the cooler rocks sink, called subduction process, creating slow, vertical currents within the mantle. This continuous movement of warmer and cooler mantle rocks produces convection cells within the mantle. These cells act like giant conveyor belts, propelling tectonic plates slowly but surely by up to 10 cm per year, about as fast as our fingernails grow. Thus the Earth, instead of appearing a solid motionless body, is a living mobile planet.

Core

Core is subdivided into outer and inner regions. The outer core constitutes the only liquid layer of the earth, while the inner core is an extremely hot, solid sphere. Both layers consist of iron and nickel, and are situated 1800-32000 miles (upper core) and 3200-3900 miles (inner core) below the surface.

Physics

Mechanics of earthquake

Background

A widely accepted theory of earthquake since the 1960s is based on the movement of tectonic plate which is a massive, irregularly shaped slab of solid rock of continental and oceanic lithosphere. Powered by forces from Earth's radioactive inner core, the tectonic plates move ponderously about at varying speeds and in different directions atop a layer of malleable asthenosphere.

Despite their tremendous weights and massive sizes, tectonic slabs manage to float about due to compositional variation of the rocks. Continental crust is composed of granitic rocks of lightweight minerals (quartz and feldspar) compared to the denser and heavier basaltic oceanic rocks.

The circulation of convection cells in the mantle is also a driving force for the movement of tectonic plates over asthenosphere.

Occurrence

The tectonic plates may collide, separate or move laterally past each other. However, sometimes two tectonic plates on grinding passed each other, can get stuck by being caught via friction of their jagged edges. When this happens, pressure builds up, until eventually the plates give way and suddenly start to slip along side each other with a jolt. The surface where they slip is called the fault plane.

When the stress at a point in the underground exceeds a critical value, failure occurs along the fault plane with a sudden displacement of the crust. The stored elastic energy is consequently released abruptly causing vibrations to travel in shockwaves (seismic waves) through Earth's crust.

These elastic waves radiating from the underground source (focus), release energy in all directions. Directly above the focus is the point called epicentre on Earth's surface from where the earthquake will be experienced most strongly. The result can be anything from a weak tremor to a full blown earthquake, depending on the amount of elastic energy released.

Earthquake and volcano

Earthquakes usually tend to spark volcanic eruptions, just like thunder following lightning. The close relationship between earthquakes and volcanic outbursts is evident from the matching of locations prone to both phenomena. The shifting of tectonic plates for an earthquake can jostle the magma reservoir beneath the fiery mountain which must be already primed to blow as a volcano.

The seismic activity of a volcano is similar to that for an earthquake, and signals that magma is moving and shifting with increasing possibility of eruption. Any tectonic movement for an earthquake with large amounts of rock moving together is similar to magma movement which is really hot rock but fluid enough to move with volcanic eruption.

Earthquake and tsunami

A tsunami is series of ocean waves with very long wavelengths (typically hundreds of kilometres) generated by large-scale disturbances of the ocean with vertical displacement of the overlying water. An earthquake can give rise to this kind of destructive ocean waves. Subduction earthquakes with giant flat slabs (tectonic plates) sliding under one another are particularly effective in generating tsunamis. When plates shoot upwards underwater, they are likely to push the water up with it. As the wail of water falls back down to sea level, it creates a monstrous wave (tsunami) that radiates outwards. The wave travels across the ocean, and becomes weaker with dissipation of energy as it travels out.

Creation of tsunami is dependent upon the depth of the earthquake. If the earthquake occurs more than 100 km below the Earth's surface, a tsunami may not, however, occur because there is not enough vertical displacement of the water.

Maths

The cause of an earthquake can be better understood if the epicentre could be traced. Experts use maths to find the epicentre and the magnitude and strength (seismic energy) of an earthquake to determine its ultimate the severity.

Monitoring earthquakes: Seismometers

Seismometers monitor the arrival times of seismic waves from an earthquake on a seismograph. The bigger the earthquake, the greater is the shaking of earth's surface. A zigzag pattern on a seismograph would correspond to varying amplitude of ground oscillations beneath the instrument. The time, locations and magnitude of an earthquake can be determined from the seismograph data.

There are two phases of the recording pattern. There is an initial pulse of small amplitudes (primary or P-waves) which reduces down slowly. This is followed by a second wave of greater amplitudes (secondary or S-waves). Based on these two waves, the location and magnitude of earthquakes can be ascertained.

Seismologists measure the interval time of S-P to find the distance from the seismometer to the epicentre. Based on the results of three different seismometers at the three different distances, specialists can find the epicentre using the method of triangulation. Thus by drawing three different circles, each around a seismometer, the point where the circles interact is taken as the epicentre of the earthquake.

Magnitude of an earthquake and Richter scale

The magnitude scale of an earthquake is a number that compares amplitudes of waves recorded on seismograms. The most commonly used magnitude scale is Richter's scale (R) developed by Charles F. Richter in 1935. It is based on the logarithm of amplitude of waves recorded on seismographs. As a result, a tenfold increase in wave amplitude would correspond to a jump in whole number of R. For example, a wave amplitude of an earthquake at R = 7 is ten times greater than an earthquake at scale 6, and would be hundred times greater than that at R= 5.

Richter's scale can accurately reflect the amount of seismic energy released by an earthquake up to about R= 6.5. This is due to "saturation" of the scale from a combination of instrument characteristics and reliance on measuring only a single, short-period peak height.

Development of a concept of magnitude scale based on surface and body waves seem to solve the "saturation" problem by extending the peak period. By definition, surface waves pass through earth's uppermost layers, while body waves travel into and through the earth.

Strength (seismic energy) of an earthquake

It is the seismic energy (E) of the quake which is responsible for knocking down buildings. An evaluation of E-values is thus more important than the magnitude/amplitude of seismic waves (R-scale) to assess the destructive power of an earthquake.

Historically, calculation of E relies on Gutenberg-Richter (G-R) energy-magnitude relation using the logarithm base 10:

$\log E \text{ (ergs)} = 1.5 * R + 11.8$

where 10^11.8 is the energy (ergs) released by a small reference earthquake based on mathematical averaging techniques. Energy from an earthquake can be converted to equivalent tons of TNT which releases releases 4.184 gigajoules or 4.184*10^16 ergs per 1 ton of TNT detonation. An earthquake at R=9.0, for example, would release energy equivalent to about 31 billion tons of TNT, or 2 million Hiroshima bombs.

Let us compare two earthquakes at R=8.5 and 5.5. The magnitude ratio is 1000, while the corresponding estimated seismic energy ratio based on the G-R relation is 31,622. It is obvious that although the magnitude/amplitude ("size") ratio is big, the ratio of seismic energy ("strength") responsible for structural damages is very high for R=8.5 compared to R= 5.5. This explains why big quakes are so much devastating than small ones.

Forecasting earthquakes

A huge responsibility is obviously associated in assessing the likelihood of earthquakes occurring. Although earquakes have been studied extensively, and their causes are well understood, prediction of such events over short time scales remains difficult.

There are two potential methods of forecasting an earthquake. In the early days, deterministic prediction was adopted to work out when, where and with what magnitude a particular earthquake would strike.

An uncertainty of the above technique led to the development probabilistic forecasting. It involved a change in strategy by calculating the odds that an earthquake above a certain size would occur within a given area and short time period. However, reliability of such prediction is not beyond any doubt.

In the wake of the L'Aquila earthquake in Italy in April 2009 after predicting, a month earlier, no such devastation forthcoming by the specialists, it is easy to realize just how complex earthquakes are and how difficult it is to predict them.

Final comments

Our fragile existence in this planet is made painfully clear by the latest terrible earthquake (9.0 R-scale) in Japan (11 March,2011) with an indication of a shift of Japan's main island by over 200 cm. It is apparent that despite our impressive control of power even for planetary destruction, we are still humbled by (when confronted with) the real power of planetary dynamics. Our achievement of harnessing the power of energy and the capability to describe regular, periodic natural phenomena may be impressive, but confident prediction of sudden planetary changes is still in its infancy.

Science of Natural Phenomena

Tsunami

The final article in the present series on natural phenomena is about tsunami. The other articles in the series are based on Earthquake, Lightning, Natural Vortex (Cyclone and Tornado) and Colourful Optics of Nature.

Tsunami throughout history is known to mankind as the most destructive natural event which can spread thousands of miles causing huge loss of life and property. Only the giant tsunamis have arguably changed history. For instance, Mediterranean tsunami struck the north shore of Crete over 3,500 years ago destroying the sophisticated Minoan civilization into a tailspin leading it to succumb to Mycenaea Greeks. In recent times, two huge tsunamis include the Indian Ocean tsunami on Boxing Day in 2004 and the Pacific Ocean tsunami in Tohoku, Japan on March11, 2011.

Meaning of "tsunami"

Tsunami is a Japanese word meaning harbour ("tsu") wave ("nami"),. It originates from fishermen's experience. While sailing for fishing, they would not encounter any unusual waves while out at sea. But on coming back to land, they were sometimes confused to find their village being devastated by huge waves at the harbour, and hence the term tsunami. An explanation of this confused experience is given later on.

Characteristics of Tsunami

Tsunami, a global phenomenon, is different from normal sea waves at the beach. The ocean waves that surfers ride are made by the blowing wind, while tsunami occurs because of geological events.

Tsunamis are associated with very long **wavelengths**, λ (crest to crest distance) in excess of 100 km and long **period, t** (time for a full wave cycle from crest through valley or trough and back to crest) from several minutes to more than an hour. Tsunami moves very fast at a **speed V** around 200 m/sec. **Wave height i.e. amplitude, L** (trough to crest) of tens of metres had been recorded in large events such as tsunami created by the landslide at Lituya Bay, Alaska, 1958.

Tsunamis are termed as "shallow water waves" which does not necessarily mean that the water is shallow. By definition, a water wave with the depth of water (**h**) much less than λ i.e. small **h/λ,** is defined as shallow, as it happens near the coast. However, in deep ocean of some kilometres in depth including the deepest part of world's oceans namely Mariana trench (in the western Pacific Ocean) ~ 11 km deep, water wave can still be defined as shallow because λ is sufficiently larger (~100km) than **h.**

Devastation by tsunami

Two recent cases

Boxing day (Indian Ocean) earthquake and tsunami

On 26 December, 2004, an undersea earthquake struck the coastal region of Indonesia making it the most powerful ever recorded in the world in the last forty years with magnitude 9.1-9.3. The movement of the sea floor produced a tsunami with waves up to 30 metres along the adjacent coast line, killing more than 240,000 people. Waves were registered all over the World Ocean which spread outward from the source (the coast of Sumatra), claiming 58,000 lives in Thailand, Sri Lanka and India within 2 hours.

Tohoku (Japan) earthquake and tsunami

On March 11, 2011, an earthquake of largest magnitude of 9.0 ever recorded in Japan produced a huge tsunami in Tohoku. The combined impacts of earthquake and tsunami left nearly 20,000 people dead, 130.000 displaced and a massive destruction along the Tohoku coast of Japan. Autopsy results showed around 96% of the victim died from drowning in tsunami waves.

Occurrence of tsunami

Tsunami could be linked either to a tidal or a seismic wave. Tides are formed only from an imbalanced extraterrestrial gravitational influence of the moon, sun and planets, while seismic waves by definition originate from earthquakes. In fact, as early as fifth century B.C., the Greek historian Thucydides suggested a connection between earthquakes and tsunamis.

Tsunamis can also develop from natural events such as cataclysmic volcanic eruption (namely Thira eruption near Crete that triggered the Minoan tsunami), or enormous landslide (Lituya Bay, Alaska in 1958 when massive rock of 40 million cubic metres plunged into water producing a huge tsunami with the largest ever recorded wave of over 500 metres high). Tsunami may also occur from unlikely events of cosmic collisions i.e. impacts of meteorite disturbing the water below. The present article focuses only on earthquake-related tsunamis.

Tsunami from earthquake

Tsunamis are mostly caused by sudden large-scale disturbances of the ocean by seafloor earthquakes which are generated along faults called subduction zones. Such faults are responsible for 90% of large earthquakes in the Pacific and Indian Oceans.

The process starts when two giant flat slabs (tectonic plates) sliding past each other get stuck by their jagged edges. The elastic energy stored in the stuck plates is enormous (equivalent to 8000 Hiroshima bombs in Tohoku earthquake). This energy is suddenly released with an enormous jolt via fracture of the plates, producing vibrations in the form of shockwaves (seismic waves). Such disturbances can cause massive vertical displacement of the overlying water to create tsunami.

It is worth pointing out that If the earthquake occurs very deep inside the Earth, say more than 100 km below the surface, a tsunami will not occur because of lack of enough vertical displacement of water to come up to the surface.

Physics of tsunami

Speed and energy/power of tsunami waves

The speed with which a shallow water wave moves is derived from **Bernoulli's theorem** as $V = \sqrt{(g * h)}$ where **g** is the acceleration due to gravity (9.8m/sec^2). In the Pacific Ocean with water depth ~5 km, a tsunami would, therefore, gather a speed ~221m/sec which is ~ 795 km/hr or close to 500 miles per hour, the speed of a jetliner. This typical speed is fast enough for a tsunami wave to travel, for instance, from Japan to the US in less than a day, or traverse the Pacific Ocean in 10-12 hours.

Tsunami's enormous kinetic energy (**K.E.**) and power (**P**) are related to the huge mass of water carried by tsunami in mid-ocean. On multiplying **h** (5000m) by the density of water (1000 kg/m^3) one obtains the volume of water as 5*10^ 6 kg per square metre of ocean surface. Tsunamis develop its high **K.E.** and **P** by moving this huge amount of water at a very high speed through the deep ocean.

Tohoku tsunami in 2011 is reported to generate ~3 petajoules i.e. 3*10^15 joules of energy which is enough to power New York City for seven days. In comparison, the total energy of the Boxing day tsunami waves in 2004 was ~ 5 megatons of TNT (20 petajoules) which is more than the total explosive energy used during the World War ll.

The rate at which a wave loses its energy is inversely proportional to λ. This would imply that giant tsunamis with very high λ can travel great trans-oceanic distances with only limited energy losses which could only arise from internal friction and viscosity due to solid-water friction.

The power **P** defined as the rate of transfer of **K.E.** is estimated in watts to be around one gigawatt i.e.10^9 watt per kilometre of shoreline.

What happens as tsunami approaches the shore

On approaching the coast, tsunami's huge **K.E.** is suddenly squeezed into a much smaller space at the shore. Based on conservation of energy, all **K.E.** is converted into potential energy. As a result, wave height is increased producing huge walls of water which start to rise up at the coast. With tsunami's speed decreasing and wave height increasing (called shoaling) near the coast, **Green's law** predicts

$L \sim 1/(h^{1/4})$. The consequence is striking. For example, if a tsunami is formed at a ocean depth **ho** with a wave height **Lo**, it will be related to the wave height **Ls** on the shore for a water depth **hs** as $Lo / Ls = [hs / ho]^{1/4}$ The above relation implies that a tsunami with a typical **Lo** of 1 m in the open ocean at **ho** $\sim$ 5 km, can end up with a final **Ls** around 8-9 m at **hs** $\sim$ 1 m near the shore. The wave height **Ls** near the shore is thus deceptively many folds higher than **Lo**, and can sometimes be as high as 30 m.

Deceptive nature of tsunami

1). On travelling along the ocean, tsunami is typically associated with relatively small **Lo** ~1 m. This would mean that at $\lambda \sim$ 100 km, steepness of the wave in the deep ocean given by $2\pi*(L/\lambda)$ is very low. Such tiny slope will manifest itself as small, harmless ripples hardly visible let alone being dangerous to the fishermen's ship on the ocean mentioned earlier. Indeed, the deep ocean is one of the safest places one can hang out when tsunamis are around.

However, on reaching the shore tsunami wave could have wave height **Ls** reaching several tens of metres, causing havoc. Fishermen's experience of devastation on returning back to their harbour was thus confusedly related to something about the harbour wave which did not show any sign of huge wave out in the sea.

2). Both **V** and **ho** decrease as tsunami approaches the shore. Based on $V = \sqrt{(g*h)}$, the travelling speed of tsunami at **hs** =10 m slows down to ~ 10 m/sec ~ 36 km/hour compared to 795km/hour in the ocean. However, the speed is still high enough to outrun a swimmer or a runner. This suddenness of speed along with huge waves at the shore would take people by surprise at the coast with devastating effect,

3). Another deception of tsunami is related to period **t** given by $t = \lambda/V$. The time required for a tsunami wave to travel in the deep ocean with λ and V around 100km and 200m/sec respectively would be 500 sec or ~ 8 min. At the shore, the depth of water and wave length if assumed to be 10m and ~ 5 km compared to 5 km and 100 km respectively in the ocean, **t** will still be 500 sec as in the deep sea.

The above estimations indicate that the typical time separating the different events of a tsunami whether it is in mid-ocean or by the coast is similar which is in minutes or even hours, than 5-20 seconds for wind-generated waves. This long break in time period can lead to tragic results by misjudging the tsunami behaviour. An initial harmless drop in water heights for normal waves can attract people to the beach which, after relatively long interval of time could follow with a devastating massive wave that cannot be outrun.

Intensity and magnitude of tsunami

Estimation of the degree of tsunami damage is primarily based on the statistical analysis of events that occurred in the past. Tsunami phenomenon is characterized, like earthquake, by intensity based on objectives of measurable parameters (energy, amplitude, period etc), and by magnitude derived from subjective descriptions reflecting scale of destructions by the incoming massive waves on the coast.

Final Comments

The massive Boxing Day superquake in 2004 and the Tohoku earthquake in 2011 stunned scientists because neither region was thought to be capable of producing a megathrust earthquake with a magnitude ~ 9. These recent incidents have prompted urgently than ever the need for scientific understanding and modelling the complicated physical phenomena associated with tsunami in order to prevent unnecessary loss of life and property. One can appreciate the difficulty in studying a full-scale physics and modelling of tsunamis that would require supercomputers and complicated software.

PART - 2

Materials and Engineering

Thoughts on Green Materials

Green living focuses on green materials that are sustainable, non-toxic and energy efficient. The basic features of green materials strategy are discussed.

There are substances in the materials world that are green in colour, namely minerals, metallic ores and alloys, chemical compounds and paints. However, the present title is a metaphor. It refers to materials (metals, plastics or composites) and their processing routes that are environmentally i.e. ecologically, (eco-) friendly and resource efficient throughout their life cycle.

Our growing awareness of issues on environmental protection is gaining increasing importance in everyday life in the field of materials science and engineering. A comprehensive review of the various features of green materials strategy is presented (1-12). Any socio-economic and legislative issues are not discussed.

Basic considerations

- Carbon footprint

All materials have an embodied energy (the energy to create them) and a carbon footprint (carbon dioxide gas released during their creation). Almost every activity of our daily life such as transportation, manufacture or anything using electricity generates a carbon footprint arising from burning a fossil fuel (coal, oil or gas) to produce energy, and releasing green house gas carbon dioxide as a by-product that makes the planet hotter. One of the aims of green materials strategy is to minimize carbon footprint.

- Life cycle

A standardised protocol (ISO 14040:1998 family of standards) refers to an analytical procedure, known as life cycle assessment (LCA) which is related to a product during all of its life cycle stages including extraction of raw materials to the end of life. One can imagine a close resemblance between the life cycles of materials and those of animals and plants. For example, extraction and synthesis of raw materials ("birth"), processing them into products to be transported and used ("maturity"), and the end of life when sent to landfill or to a recycling facility ("death") can be considered as various phases in the life of a material.

LCA methodology, although expensive and approximate in nature, analyzes and can pin down the stage that consumes most of the resources and generates more carbon dioxide than all others put together.

- Environmental effect

Although the life cycle stages for different materials can be described in identical terms, the impact of environment on them can differ significantly. For metals, the mining, extraction and refining stages ("birth") are often very energy intensive, and cause pollution of air (fossil fuel-related emission of carbon dioxide), water and soil. Based on data (6), eco-toxicity and global warming in the "birth" phase of base metals like magnesium, iron, steel and copper are much less than those for the precious metals such as gold, silver and platinum. For biological materials, however, the "birth" stage relates to growth which is relatively emission/pollution free.

Environmental impacts in the "maturity" phase depend on the specific application of the material in terms of product lifetime. This could be problematic, as materials are incorporated into products and it is the products that provide functionality. However, alternative designs using different materials can provide an indication of the environmental implication of material choice. For example, in transportation equipment, lightweight materials such as aluminium, magnesium or fibre composites can provide substantial fuel savings. In homes, the use of insulating material would provide higher energy savings.

Usefulness of green materials strategy

The basic issues of the strategy should include choices of materials and manufacturing processes that would affect a product's life cycle. Wider product design objectives with an eco-friendly strategic approach to materials decisions are essential. It is important to discuss why and how materials strategies could be improved in manufacturing organizations who need to do much more for a better environment than simply delivering products that may excel at their required function. They need to address strategic drivers such as cost, environmental legislation and global manufacturing.

Also, business successes are no longer guaranteed simply by standing still without any innovative research on products. A systematic approach to materials strategy should embrace consistent consideration of both functional performance and life cycle cost implications. The goal is, as always, to optimise product performance against cost effectiveness.

Increasing strategic pressures should encourage engineers to modify product design objectives which would ultimately lead to minimizing cost, rendering the product more attractive and portable, or making them smaller and lighter, reducing environmental impact, and meeting health and safety regulations. The industrial implementation of green materials is important to manufacturers of parts such as automotive and aerospace components, domestic appliances, and parts for heating, ventilation and air conditioning.

Decisions on materials play a pivotal role in achieving product design objectives. Some of the property data requirements should include strength, resistance to metallurgical fatigue and creep, thermal conductivity, thermal stability, chemical resistance etc of the materials. With new eco regulations, it is a challenging task during the product design stage to match functional requirements and the constraints on options of materials imposed by legislation or other regulatory issues.

Application of green materials strategy

*   Materials selection

Rational selection of materials to meet environmental objectives starts by identifying the phase of the product life that causes great concern: production, manufacture, transport, use or disposal. The principles of materials selection involves seeking the best match between design requirements and the properties of the materials that might be used to make the product. Most often this choice is based on an engineer's experience, supplier's recommendation or simply on what was used before. Although such approaches work in many cases, they inhibit innovation, and thereby do not allow any systematic exploration of all possibilities that could deliver better products.

Software tools containing materials databases on the functional requirements of a product can identify a candidate or substitute material by comparing composition or property profiles. The Cambridge Engineering Selector(CES):2010 software developed by Granta Design, U.K. (5,7-9) simplifies the material selection criteria with a broader objective of minimizing cost.

Charts on material property are required as a guide to materials selection in order to minimize mass, total embodied energy and thermal losses. Such consideration should include charts of density and strength against Young's modulus and embodied energy, and chart of thermal conductivity against thermal diffusivity (1-5, 7-9). However, methods involving simple property or composition comparison cannot identify the possibility of supplying a cheaper material of different composition and property profiles if changes to design were also made. Similarly, simple comparisons may not lead to an evaluation of trade-offs, such as those between cost and weight.

- Minimizing cost

Material performance for a specific application is often governed by a combination of properties. The functional properties must be combined with cost effectiveness whether it is financial cost, environmental cost of energy and carbon footprint, or the packaging costs of weight or volume. One would be able to generate complete and auditable answers to real business questions simply by enabling systematic selection of materials via a software tool that covers cost per unit function. Cost reduction strategies for materials could have several options, namely taking out cost at the production stage, materials rationalisation by using fewer materials and suppliers, or by designing out cost for next generation of products. These strategies can be improved by assembling data on likely candidate materials (1-5).

- Eco-auditing

An eco-audit is a fast initial assessment of the life of a material that carries the demand for embodied energy or carries the biggest burden of carbon footprint. The main purpose of an eco-audit is comparison, allowing alternative design choices to be explored rapidly (1-5, 11). There are challenges in integrating eco-design into product development, and managing restricted and hazardous substances.

The eco-audit procedure is derived from the LCA methodology in terms of structure, framework and requirements. Computer-aided eco-auditing can be carried out with a well established LCA database using CES eco-audit software (5). The procedure implements input data on the bill of materials, manufacturing choice, transport needs, duty cycle (detailing the energy and intensity of use) and disposal route. The outputs, on the other hand, quite logically include data on energy consumption and carbon footprint of each phase of life. Case studies of eco-auditing have been reported on products such as the electric kettle, coffee maker, auto bumper and family car using data sheets (1-3) or software packages (5,11).

Conclusive remarks

A significant challenge for materials scientists and engineers today is to guide decisions that would alleviate/eliminate adverse eco-impacts. A total eco-plan should incorporate the production method, use of products and the the materials from which they are made of, because these issues often involve significant energy inputs that would affect the climate via green house gas emission. The complex situation of frequent conflict between minimizing environmental impact and minimizing cost appears to be an ongoing task for the future.

References

1. M.F. Ashby: Materials selection in mechanical design, 3rd ed., 2005
2. M.F. Ashby & K. Johnson: Materials and Design, 2002
3. M.F. Ashby: Materials and the environment, 2009
4. M.F. Ashby et.al.: Materials: engineering, science, processing and design, 2007

Ref. 1-4 are published by Butterworth-Heineman, Oxford

5. www.grantadesign.com
6. M.H. Classen et.al.: Life Cycle Inventories of Metals, Dubendorf: Swiss Centre for Life-Cycle Inventories, 2007
7. www.mdmc.net
8. www.grantadesign.com/products/mi/emo.htm
9. www.grantadesign.com/products/data/
10. D. Cebon & M.F. Ashby: MRS Bulletin 31 (12) 1004; 2006
11. B. De Benedetti et.al. Materials Trans. 51(5)832,2010
12. S.J. Davis & K. Caldeira, Proc. National Academy Sci.,107 (12) 5687, 2010

Metallurgical Ingenuity of Ancient Humans

Metallurgy has been a hallmark of mankind's early civilization. The incredible metallurgical expertise of ancient humans is reviewed.

Metallurgy has played a pivotal role in every known civilization. From coins to weapons, tools to decorations, no substance has ever been as important as metal for the advancement of mankind from its earliest known times.

Metallurgy can be defined as the science and engineering of metal extraction from ores/minerals for subsequent use. It also covers processes for metal artefacts. The first evidence of human metallurgy in Western Europe dates back to 5th and 6th millennium BC (excavations at Cerro Virtud in Almeria, Spain, in 1994).

The mysteries of ancient activities in handling metals are subject to much curiosity, marvel and academic debates. The present review describes the fascinating history of the incredible metallurgical activities of the ancient humans.

Definition of ancient time

We do not know when man first appeared on earth. Ancient time refers to combined Prehistoric Age (i.e., before developing the art of writing) and Historic Age. The latter era began in the valleys of the Nile and Euphrates (Saudi Arabia), with evidence of written records that were made at least 4,000 or 5,000 years before Christ. In the present context, activities over four archaeological epochs, namely Stone (8000-4000 BC), Copper (4000-3150 BC), Bronze (3150-1200 BC) and Iron Ages (1200-580 BC) are considered. The ages are named after the dominant metals of the day. The ranges of the ages overlap each other, just as in modern times when the Age of Steam runs into the Age of Electricity.

Achievements in mining metallic ores

Earth's surface

In metallurgy, mining of ores is an integral part of extraction of metals. The oldest mine, Lion Cave in Swaziland, is radiocarbon dated to be ~43,000 years old. It is a mine of red ochre, which is an iron ore (hematite), used for religious occasions. Stone hammers, crude antler picks of bone or wood, etc. were used for minimg metals close to Earth's surface. This basic method was obviously too difficult to apply for extracting metals that resided deep in the mountain.

Inside mountain

In the absence of explosives used in modern-day practice to blast mountain rocks, ancient humans around 5000 BC came up with intriguing ideas of hydraulic mining and fire-setting. Thus, once an ore vein was discovered, numerous aqueducts were built to supply water to a tank in the mine head. Water was sometimes lifted by reverse overshot water wheels. Once a full tank was opened, a flood of water rushed down the hill, exposing the bedrock underneath. The rock face was next heated by setting fire to it, followed by quenching in a stream of water. The thermal shock shattered the rock into pieces for easy transport to a suitable site for extraction of metals. Open cast mining was adopted for safe dissipation of the smoke and fumes. Deep shafts are reported being cut as early as 4000 BC into the hillside at Rudna Glava in the Balkans to mine copper ore.

Achievements in metal extraction

The principles of metal extraction in ancient time are the same as used in the present day.

Panning

It is the oldest (~first millenium BC) and simplest method of mining heavy metals, such as gold, from riverbanks. Thus, once a suitable deposit of say gold is located by streams, some sediment is scooped from it into a pan where it is gently agitated in water. Light materials would spill out of the pan, leaving gold to sink to the bottom of the sediment for collection by the panner.

Smelting

A campfire theory appears to be most popular for the origin of smelting. Accordingly, ores were accidentally used to build stone closures around cooking fires when people noticed new materials appearing from ashes (due to smelting of ores). The process may have worked for lead and mercury ores, which are easily smelted at low temperatures. However, campfires are about 700 C short of the temperature needed for smelting base metals like copper, which may have been achieved in pottery kilns developed ~6000 BC using charcoal fire.

In smelting, metals are produced from oxide or carbonate ores by high-temperature thermal reduction using a reducing substance, such as charcoal. Evidence of earliest smelting of copper was found in Turkey around 6000 BC compared to 5000 and 4000 BC in Spain and Jordan when the process was operated at low temperature in poor reducing conditions with air supplied by blowpipes to oxidise sulphide ores. Major developments using small furnaces with bellows to blow air occurred during 3000 BC as evidenced in Chrysokamino, Crete. A process of co-smelting reduction is believed to have occurred between the 2nd and 1st millenium BC at Nil Kham Haeng, Thailand, when copper was recovered via simultaneous interaction between mixed copper ores of oxide and sulphide. Smelting of iron dated back much later, ~930 BC.

Cupellation

The process of cupellation (of silver) started in the early Bronze Age when lead sulphide ore (galena) containing silver was heated at high temperature (900-1000 C) in a cupel/hearth with airflow for oxidation. Lead and other base metals were transformed into oxides to be absorbed into the cupel walls by capillary action, leaving behind silver in the cupel.

Achievements in metal processing

Hammering and annealing

Neolithic cultures used naturally occuring pliable native copper from 8th million BC, which could be hammered into complex shapes. The ancient craftsman was, however, puzzled to find out that a metal once quite ductile behaved differently by not responding to additional hammering once it was hard. He, however, noticed that the metal became soft again when exposed to fire. This led to the discovery of an annealing process that removes stresses in the crystal lattice caused by cold work (hammering). Metallogrphy of native copper items from 6400 BC showed recrystallized structure as indicative of annealing. Hammered foils and gilded copper from the Andean region dates back to 1400-1100 BC.

Quenching

Much of the history of quenching developed out of mysticism and empirical experimentation. The period between ~400 BC and ~1500 AD is shrouded in mystery, perhaps due to the desire of blacksmiths to protect this unique process. One of the first mentions of quenching is from Homer (~800 BC) when a hot axe blade was plunged into cold water to make it strong.

Surface carburizing

This process was used ~1000 BC in Egypt and Europe to improve wear resistance of steel. A carbon-rich surface layer was produced on steel tools left in a red hot bed of charcoal. The hot steel thus prepared have a dramatic increase in surface hardness hence wear resistance (dut to martesite formation) on quenching in water. The development of the technique could be due to super skill of the then metalsmiths or a fortuitous incident. Carburizing today also aims at the same goal of wear-resistant surface and a soft but tough core in steel.

Guilding

The surface of chisels from a 3rd Dynasty grave (~2686 BC), initially believed to be gold, had coppery colour underneath blistered areas. Examination of samples revealed gold-rich surface merging into the chisel body of baser alloy of copper, gold and silver. Guilding is a process of covering an item with a thin layer (~1 micrometre) of gold, which is believed to be hammered onto the surface of the item in ancient time. Hammered foils and guilded copper dated 1200-1000 BC were also found in Mina Perdida in Peru.

Lost-wax process

Investment casting using this method is one of the oldest and also the most modern metallurgical technique for making intricate shapes. Metals are thus handled by shaping them into hollow, solid forms by casting them in wax moulds. The earliest known lost-wax casting is an Indian bronze figurine dating back ~5000 BC. Classical Greek and Roman sculptures were produced using the technique in the Aegean during the Bronze Age. The Egyptians used the process from mid-3rd millenium BC as shown by early Dynastic bracelets and gold jewellery. The Aztec goldsmiths in pre-Columbian Mexico (500-200 BC) created their elaborate jewellery by this technique.

Concluding remarks

Metals have defined the technological and economic character of the ancient life. Metallurgy in the present context (i.e., end of first millenium BC) may be rudimentary, but it got the job done nevertheless. With the advent of modern analytical tools, including carbon dating method, we are beginning to understand the ramification of man's first sophisticated uses of metals.

Materials and Technology: Past Period

Part 1

Metallurgy of the ancient and medieval periods is reviewed. A similar review is presented for modern time with an outlook for the future in Part 2 and 3.

Prehistoric Pottery

Materials have enabled advancement of mankind from its earliest beginnings. The progress of cave-dwelling humans to today's modern society owes much to Mother Natures' treasury of metal and its alloys. Our current use of metals is the culmination of a long path of development extending over approximately 6500 years. In fact, the first evidence of human metallurgy was found in Serbia dated around 5th to 6th millennium BC. The present article provides a comprehensive review of the ingenious usages of materials and technologies in ancient and medieval times.

Ancient Period (up to the end of first millennium BC)

The different responses of materials to varied situations enabled ancient humans to exploit and benefit from materials' distinctive characteristics in everyday life.

Ceramics

Ceramics are the most durable of all materials. Shards of pottery dating back 18,000 years were unearthed in a cave in China. Evidence showed that humans were making containers out of fired clay. Ceramic pots and ornaments survive from 5000 BC. Prehistoric pottery was handmade rather than wheel-turned. By the 1st century BC, wheel-made pottery including decorated tableware was available during the early Roman period (post 500 BC). Pottery in ancient times was a source for artistic expression, and in the form of pots as tools for cooking, serving and storing food. Roman cement as ceramic is still in use to bond walls of villas

Glasses

Natural glass has existed since the beginning of time—the oldest fragments of glass vases dated back to the 16th century BC. Glass was formed when certain types of rocks melt as a result of high-temperature phenomena such as volcanic eruptions or lightning strikes followed by rapid cooling of the molten rock. Stone-age man is believed to use such rock, known as obsidian rock, to make cutting tools. The earliest man-made glass beads dated back to ~ 3500BC.

Materials timeline

This is best described in four archaeological epochs named after the dominant material of the day:

1). Stone Age (8000-4000 BC)—weapons and instruments made of bone, flint, quartz and other kinds of stone
2). Copper Age (4000-3150 BC)—associated with tools, weapons and ornaments
3). Bronze Age (3150-1200 BC)—bronze compared to copper, allowed superior arms and cutting instruments, and
4). Iron Age (1200 BC-1500 AD)—supersedes the demand for bronze in making arms, axes, knives, swords etc, although bronze was still being used in this period for ornaments and also for the handles of swords and other arms but never for the blades.

Native metals

Gold (6000 BC), copper (4200 BC) and silver (4000 BC) were the first known metals to be discovered in nature. These native metals were appreciated for their ornamental and utilitarian value. Gold was one of the first metals being exploited with its superior chemical stability than the others. All three metals being ductile, were used for making ornaments.

Work hardening of metals

It was realized around 5000 BC that native metals could be beaten to complex shape and become hard. Thousands of uniformly "pressed" gold beads dated around 3000 BC were discovered in Bulgarian "Valley of Kings". Fashion wares of copper, lead, zinc, iron, gold, tin and bronze (alloy of copper and tin) were found In Metsamor, Armenia that dated back around 2500 BC. The history of Andean region reveals hammered foils and gilded copper being preserved dating back to 1400 to 1100 BC. However, copper even in worked state, was not hard enough for wear resistance. As a result, copper weapons and tools were easily blunted. It was around 3000 BC, a new copper-tin alloy technology (Bronze period) opened up, probably by accidental inclusion of a tin-based mineral (cassiterite) in a copper ore. Tin provided bronze with hardness unmatched by copper alone, and thus allowed ancient humans to make superior tools and weapons.

Technology

Perhaps the earliest known use of technology was in the Stone Age when the first knife or shovel was made from a piece of stone or volcanic rock (obsidian). Materials in the Bronze and Iron Ages were the first ones to be engineered by changing them to fit what they needed to do, rather than just letting their natural properties determine what they could be used for. The principles of metal extraction technology in ancient times are the same as in the present day, namely panning for gold (~1st millennium BC), smelting of copper ore (3000-6000 BC) and cupellation of silver(early Bronze Age). Several techniques such as hammering and annealing, quenching and surface carburizing were used to process metals, mostly with the aim of making them harder and more wear resistant. Also, relics of guilding and lost-wax process indicate the artistic achievements of ancient humans.

Medieval Time (up to the middle of second millennium AD)

The medieval period or the Middle Ages followed the fall of the Western Roman Empire in 476 AD and preceded the early Modern era. A look at the history of metals would indicate that it was only the Iron Age that continued, leaving a total vacuum of discovery of new metals. Invention of materials between the 5th and 14th century was limited, except for developing materials like silk which was more quantitative than qualitative in character. However, there had been several technological innovations during the medieval period.

Technology

Harnessing mechanical power

The use of mechanical power was the primary development technology. Prior to the Industrial Revolution of the 18th Century, spectacular inventions had been the watermill, windmill and wheeled plough. The Greeks invented the two major components of watermill, namely the water wheel and toothed gearing whose first technical description dated back to 40/10 BC. Greeks with Romans operated various types of watermills demonstrating for the first time that power can be harnessed by man, and not necessarily needed to be generated by animals or humans. With the development of windmill, it was possible to harness the energy of the wind effectively. The windmill was created to replace animal power in grinding grains. Regarding origins of these innovation, it is argued that a fair number of medieval inventions including windmill were not inherited from the Greco-Roman world, but borrowed from the east, namely China and Persia.

Building

The medieval era was fraught with violence because of the threat of war (Crusades) or land disputes. One of the greatest protection skills of the day was to build up castles that would act as power bases for fortification. The development of lifting appliances using a simple system of block and tackle and cranes with a treadmill, stimulated the growth of buildings such as castles and churches.

The building materials that were used varied through history. Wood being cheap and simple to handle, was used to construct most castles in the first millennium BC until 1066 AD. However, it fell into disuse because of flammability. As a result, inflammable and much stronger ceramics such as stone soon became popular. However, it was more expensive and took longer to construct buildings in stones. With cathedrals as proofs, medieval craftsmen displayed artistic genius, skill, and daring. However, quality of buildings in that era was generally poor, requiring frequent repair, replacement or reconstruction.

Weaponry

Threat of war led to the evolution of weaponry and military skill. Medieval weapons included sword and battle axe (forged from steel and sharpened), mace (developed from a steel ball on a wooden handle attached to a spiked steel war club), dagger (wood or animal bones), ballista (similar to crossbows) etc. Damascus steel was used in making swords which are characterized by distinctive patterns of banding and mottling—reminiscent of flowing water. The blades thus made were reported to be tough and resistant to shattering. Regarding medieval shields, their constructions depended on the style of combat and armour worn. Materials for shield included wood, animal hides, woven reeds or split-resistant timber reinforced with metal.

Final Comments

Man's technological ascent began in earnest in prehistoric time, and innovations continued through the Middle Ages in the first few hundred years after the fall of the Roman Empire. Advancements in technology in this era saw the use of simple machines such as windmill and watermill.

Materials Development: Present Time and Future

Part 2

Following Part 1, modern and future materials are reviewed. Technological innovation for the same period is a vast topic and will be reviewed in Part 3.

Human civilization owes its development to progresses in materials and technology. In the modern era, we have vast numbers of tailored materials to make use of. Like various Ages in ancient times, we can consider ourselves living in Materials Age. Following the philosophy of "one faces the future with one's past", progress in materials of the present time is reviewed, followed by a brief account of futuristic materials.

Development of modern materials:

Plastics

Polymers or plastics are the chemists' contribution to the materials world. They are generally light organic compounds of carbon and hydrogen, and consist of large molecular structures. Plastics play an important part in everyday life. Their versatility allows them to be used in everything from shoes to ships and from food packaging to fighter jets.

Silicon-based material

The last forty years of rapid development of electronics technology such as microelectronics, can be described as the Silicon Age. There would have been no information superhighway as we know it if there was no development of high-purity silicon and silicon-based integrated circuits. Based on total internal reflection phenomenon, fibre optic cables have revolutionised long-distance phone calls, cable TV and the internet by carrying out digital information via transmission of light signals. These cables are long strands of optically pure glass (silica) as thin as human hair.

Composites

These are complex materials in which two or more structurally complementary substances combine to provide properties not present in the individual component. Composite technology is not new. Straw-reinforced mud brick (adobe) is one of the earliest architectural composite materials still in use in parts of Africa and Asia. Steel-reinforced concrete for building construction appeared around 1850. Material developments in the 20th century include fibreglass where glass and polymer are combined. Other new combinations include ceramic fibres in metal or polymer matrix. The fibres carry the mechanical loads, whereas the matrix material transmits load to the fibres and provides ductility and toughness.

Metals and alloys

Our everyday usages of metals and alloys range from cars, airplanes, computer chips, mobile phones, refrigerators, microwave ovens, TVs to biomedical devices for replacement of joints and limbs. The various properties discussed below are utilized for specific applications.

Thermal properties

Besides pure metals, just about every single piece of metallic stuff in our daily life is an alloy of some kind, such as steel and cast iron (alloy of iron and carbon), brass (copper and zinc), stainless steel (nickel, chromium and iron) etc. Modern research in alloys is directed towards improving their performance with subsequent use which no one could have imagined possible a few years ago. For example, the development of nickel-based "Superalloy" with exceptionally high temperature strength (capable of operating at as high as 2000 F), has made it popular for use as turbine blade in superhot turbine areas of modern jet engines, making present day transportation possible.

Mechanical properties

There are some crucial modern applications where a material should neither undergo excessive change in shape, nor break down when mechanically loaded. Examples include load-bearing structures such as materials for aircraft wings, jet engine or bridges. It may be, however, a secondary consideration in cases like in the communication industry where the electrical properties are of first concern, although the need for a device or optical fibre not to fracture when in service, would be a close second.

Fail-safe consideration

Unlike in the past, it is essential that materials on locations do not fail, since any repair or replacement is either very expensive, extremely difficult or impossible. Such locations include deep-sea oil pipelines, space stations, satellites and healthcare biomaterials within our bodies.

Development of future materials

New materials have been revolutionising our daily life for over three decades. The possibilities of developing future materials would depend on our imagination and ability to make and test them. Several futuristic materials are briefly discussed.

Superconductor

A striking development in the 20th century was in 1911 when mercury and lead cooled to 4.2 degrees K (-269 C) suddenly showed no electrical resistance. However, the temperature was too low for such materials to be of any use. In the mid-1980s, a complex oxide of barium, lanthanum and copper was found to be superconducting at 30 K. This prompted in 1987 to develop superconductor that worked at liquid nitrogen temperature (98 K), followed by invention of ceramic superconductor in 1993 that worked at 138 K. Effort to develop purely metallic superconductor is in progress. Practical applications have so far been limited to a few areas only, such as elimination of friction in trainstransportation vehicles, bio magnetic technology as depth sensor for medical equipment, and acceleration of sub-atomic particles nearly to the speed of light in Super Collider.

Graphene

In 2004, graphene, a new form of carbon, was discovered which consisted of carbon atoms of just one atom thick. It displayed unprecedented properties of being one of the strongest materials ever tested. Sheets of graphene have been used to make the fastest transistors ever made. However, commercial viability of graphene is yet to be developed because only one small sheet can be currently produced at a time.

There is a possibility of partial to complete replacement of plastics by incorporating graphene into any product that uses plastic. Composites thus produced are expected to be stronger, lighter and more environmentally friendly than their plastic counterpart. Applications could be in aircraft parts, car parts, sports and household goods. This would be a great achievement in the plastic world where the major raw material to manufacture plastics, oil, is a limited resource.

Fullerene

Besides existing as diamond and graphite, carbon also exists as fullerene (buckyball) molecule of 60 carbon atoms arranged in pentagons and hexagons. Fullerene has no crystalline structure i.e. amorphous, with a high bulk modulus of 491 gigapascals (GPa) compared to diamond's 442 GPa. Work is in progress for its application in various areas such as organic photovoltaics, polymer electronics, antioxidants, nanotechnology, and so on.

Nanotubes

Carbon nanotubes are long chains of carbon atoms held together by the strongest bond ever achievable in chemistry. Among their numerous remarkable physical properties is ballistic electron transport—making them ideal for electronics. Their strength (48,000 kN.m/kg) is about 300 times stronger than steel and would make them capable of building space elevator.

Amorphous metals

Amorphous metals also known as metallic glasses produced by very rapid cooling of molten alloys, consist of a disordered atomic structure. They can be twice as strong as steel. Because of their disordered structure, they can disperse impact energy more effectively than a crystalline metal which has points of weakness. Such materials can thus be useful in the military's next generation of armour or aircraft components which must deform to absorb the energy of collision, but must also become gradually stiff during the crunch. Their electronic properties are claimed to improve the efficiency of power grids by as much as 40%.

Metal foam

Metal foam is a cellular structure of a solid metal (aluminium or nickel) containing a large volume fraction of gas-filled pores or foaming agent (powdered titanium hydride). It is a very strong substance that is relatively light with 75-95% empty space. Because of favourable strength to weight ratio, metal foams are proposed for construction of stronger buildings, space colonies, orthopaedic applications and automobile uses. Composite foam of hollow steel spheres surrounded by aluminium is being investigated as a possible building material.

Metamaterials

Metamaterial refers to any material that gains its properties from its certain periodic patterns and shape rather than composition.

These artificial materials engineered to have properties that may not be found in nature, guide light around an object, rather than reflect or refract the light. Metamaterials still in the process of development have been used to create microwave invisibility cloaks, 2D invisibility cloaks and other unusual optical properties. Mother of pearl (organic-inorganic composite) gets its rainbow colour from metamaterials of biological origin.

Final Comments

Developments of metals like bronze and iron enabled advances in civilization thousands of years ago. This synergy continues today, for example in fibre optics which have created the World Wide Web, and among others, the development of biomaterials that mimic living tissues. One would hope that with good scientific knowledge and ingenuity we will be able to strive forward in grand style making startling changes in our lives in the 21st century.

Engineering Technology: Present Period and Future

Part 3

In human history, engineering has driven the advance of civilization (Parts 1 & 2). Modern and future engineering technologies are reviewed in Part 3.

The origin of engineering goes back to the very beginning of human civilization when tools from stones and bones were first created. Products and processes that enhance the joy of living remain a top priority of engineering innovation, since the day of the taming of fire (prehistoric time) and the invention of wheel (medieval period).

From the metallurgists (who ended the Stone Age) to the aircraft builders (who brought the people of the world closer), the past witnessed many marvels of engineering prowess. A comprehensive review of the vast field of modern engineering technology since the 19th century is presented with an outlook for the future.

Defining Engineering Technology

A look at our mundane household requirements such as internet, printer, telephone, home entertainment, etc, would reveal very little that is truly "natural", and not created by engineering technology. Present day engineering technology is devoted to the implementation and extension of existing technology for the benefit of humanity. Engineering technology is based upon similar knowledge as engineering, but focuses mainly on manufacture and field service.

Engineering Timeline

The concept of engineering since medieval times with the inventions of pulley, lever and wheel is consistent with the modern definition of engineering. The history of modern engineering in the last two hundred years can be grouped into two overlapping phases:

1. The Industrial Revolution of the 18th continued through the 19th century up until the first half of the 20th century when electricity, telecommunications, cars and airplanes were developed, and
2. The development of information technology in the post World War ll period via combination of microelectronics, computers and telecommunications.

Modern Engineering Technology (up to the beginning of 21st century)

Technology is striding forward at such a fast pace in the last 200 years that what is considered a marvellous novelty turns into a mundane common object as the world goes for the next level of modern thing.

19th Century

The invention of useable electricity, steel and petroleum products during the19th century lead to improved communication and growth of railways and steam ships. The best inventions of 19th century engineering technology can be classified as

Steam-powered ships, trains and cars and internal combustion engines (transportation)

*   Telephone, telegraph, radio and television (communication)
*   Electric motor, incandescent light bulb (harnessing electricity)
*   Rifles, Colt revolver, machine guns and silencers (firearms), and

- Skyscrapers and large cities (construction).

20th Century

The late 20th century is undoubtedly the Age of Technology rivalling the Industrial Revolution in its impact on civilization. The wide-spread distribution of electricity and clean water, car and airplanes, radio and TV, spacecraft and lasers, antibiotics and medical imaging, computers, cell phones and Internet are some of the features which improved the standard of human life in the 20th century. Computers were improved by utilizing miniaturized transistors and integrated circuits.

Outer space was explored with satellite which was later used for telecommunication. By the middle of 20th century, humans had enough mastery of technology to be able to leave the Earth for the first time and explore space (man landing on the moon in1969). With so many new developments and discoveries, it is hard to narrow down to a few. However, the following items among a plethora of inventions are singled out as important engineering inventions in the 20th Century:

- Internet (1960-1970), Modern Internet (1990-2000)
- Personal Computer-Microprocessor (1971), Windows (1985)
- Lasers (1958) and Radio Space age (end of the 60s)
- Silicon single crystals grown for semiconductors (1960)
- Television (1950) HD TV
- Airplanes (1903)
- Fiber Optic(1955)
- Space Rocket (Explorer, 1958)
- Discovery of Nuclear fission which led to developing nuclear weapons and nuclear power
- Genetic Engineering which alters the traits of living organisms by manipulating information encoded in their DNA. It can increase crop yield or resistance to desease by producing new enzyme or protein, and
- Nanotechnology manipulates materials on a nanometre scale (one billionth of a metre), and is based on thorough control of

the structure of matter to produce goods such as microprocessors, batteries, computer displays, paints and cosmetics of better quality at lower cost and a cleaner environment.

First Decade of the 21st Century

The main technology in the early 21st century is based on electronics such as Broadband Internet access. Other developments include:

- iPhone (2007)—an internet-enabled multimedia mobile phone with multi-touch screen and virtual keyboard. iPod (2010)—a portable digital media player which can play videos or audio files such as MP3
- Electric eyes—retinal implant (2009)—a microelectronic chip is developed for implanting into the patient's eyeball (suffering from age-related macular degeneration and blindness). A pair of glasses equipped with a tiny camera will transmit information to that microchip which in turn will pass along recognisable visual information to the human brain to allow the patient to recognise objects, and
- Sixth sense(2010)—a wearable gestural computer interface that lets hand gestures interact with digital information. It comprises of a pocket projector, a mirror and a camera which allows one to interact with objects in a way never before possible. Still a prototype, but provides a real possibility of taking a picture simply by framing scenery with hands.

Future Outlook

No one knows for certain what lies ahead since it is virtually impossible to predict future inventions. Tomorrow's world is undoubtedly going to be very much different from today's. For example, no one could have predicted five hundred years ago that we will be now flying around in planes.

- Based on today's evidence, one can figure out what is coming next and what effect it will have. Some examples are expressed below:

- Developing affordable solar energy technology that would convert and store the power of sunshine at a cost compatible with fossil fuels.
- Providing energy for commercial power generation with controlled but sustained nuclear fusion.
- Reverse-engineering of the brain to unlock the secrets of brain function, heal human diseases and advance the field of computerized artificial intelligence which in turn should enable automated diagnosis and prescriptions for treatment.
- Securing cyberspace by protecting the global infrastructure from identity theft, viruses, etc.
- Enhancing virtual reality by creating imaginative environments for education and entertainment using computer technology.
- NASA's future space missions include communicating data from a spacecraft with lasers instead of current use of radio waves, and developing atomic clock based on a mercury atom for testing ultra-precise timing in deep space.
- Progressing genetic engineering with genetically modified plants, animals and micro-organisms and genetic medicine.
- Developing nanotechnology that would combine engineering principles with known scientific facts for future technological possibilities for biomedical applications and manufacturing goods from nearly flawless, superstrong materials at low cost, and
- Technological development to manufacture metamaterials on a dimension smaller than the nanometer wavelength of light so that light would be blocked to create cloaking devices invisible to the naked eye or radar (useful in military applications).

Bioprinting, virtual reality and quantum computing are some of the other prospective technologies of the future.

Final Comments

Since ancient time, technology has been used to change the world by altering the manner of our existence with improved quality of life. For the future, information technology is expected to continue changing the ways we work, learn, and interact with each other.

Modern engineers are being viewed as capable of analyzing today's problems, developing tomorrow's solutions, and fixing yesterday's errors. In the 21st Century, one can look forward to exciting technological possibilities that lie ahead.